THE COMPLETE GUIDE TO

Beekeeping

THE COMPLETE GUIDE TO
Beekeeping

NEW, REVISED EDITION

ROGER A. MORSE

E. P. DUTTON | NEW YORK

For information contact: E.P. Dutton, 2 Park Avenue,
New York, N.Y. 10016

Library of Congress Catalog Card Number: 79-93330

ISBN: 0-87690-126-7 (cloth)
ISBN: 0-525-93105-8 (paper)

Published simultaneously in Canada by Clarke, Irwin & Company
Limited, Toronto and Vancouver

10 9 8 7 6 5 4 3 2

Contents

I

Bees and Beekeeping

Man has not changed the honeybee. The bee we use today is the same as it was thousands of years ago; however, we have learned much about the biology of the honeybee and how to keep it in a man-made hive. A successful beekeeper applies man's accumulated knowledge of bee biology to assist the bee in honey making.

The honeybee is also an important animal to those interested in biology. In science we want to know more about the evolution of animals. How did social animals come into existence? How do honeybees communicate? What are the forces that hold the colony together? The answers to these questions are not the subject of this book and are discussed only in brief. However, before anyone, scientist or amateur biologist, can use the honeybee experimentally, he must understand its elementary biology.

The goal in beekeeping should be honey production; this book is concerned with that goal. Only a beekeeper who produces as much honey as possible thoroughly understands bees, beekeeping and bee biology. Producing a crop of honey, liquid or comb, brings into play many factors: wintering; spring build-up; good queens; the importance of food to the colony;

major and minor honey flows; good combs; pests and diseases; bee space; and the proper equipment for keeping bees.

Unlike many animals, bees do not need daily attention. Yet, at certain times of the year specific operations must be performed. Thus a knowledge of bee biology is necessary.

Probably more people, perhaps too many, would keep bees if they did not sting. In times of danger any animal must defend itself. Bees sting to defend their nest. Their only other alternative is to flee (abscond); this occurs only rarely with the honeybee used in North America. Bee stings, or at least most stings, can be avoided. Just as with good bee management, avoiding bee stings requires a knowledge of bee behavior. Still, any beekeeper expects to be stung occasionally.

A honeybee is a social insect. An individual worker honeybee, drone, or queen cannot live alone. A minimum of about 200 worker bees and a queen are required for social order to exist. Even this number is too small if the colony is not warmed artificially. For practical purposes a colony should have at least four thousand to five thousand worker bees (about one pound) and a queen. Drones (males) are present during only the late spring and summer, sometimes in the fall. The sole function of the drone is to mate with the queen and drones need not be present to have a successful colony.

In the case of solitary, or subsocial, bees, the female selects and builds the nest, gathers the food, feeds the young, guards or protect the nest, and sometimes lives long enough to see her offspring. Not so with honeybees; they have a division of labor. The honeybee queen lays all the eggs and provides the colony with certain chemical substances that control social order. A young worker bee, after it emerges from its cell, first cleans cells and later feeds larvae, deposited in the cells as eggs by the queen, before graduating to a series of other hive or "household" tasks. The worker bee's last duty as a house bee is to be a guard; thereafter it works only in the field. While division of labor within a colony has some flexibility, no individual is equipped to do all of the required tasks at one time. A division of labor requires numbers. Added to the routine tasks of feeding, cleaning and foraging is that of hive temperature and humidity control. This, too, requires a large number of worker bees. If the hive temperature falls too low, the brood will be chilled and die.

Bees are also important to the life of flowering plants. Unlike many insects and animals, honeybees live exclusively on pollen and nectar. Thus their lives are closely entwined with the lives of flowering plants. Plants produce pollen and nectar to attract honeybees so that the flowers can be cross-pollinated. In return, honeybees carry pollen from one flower to another. Without flowering plants there could be no bees; without bees certain flowering plants could not exist. Scientists have a reasonable understanding of how this case of interdependence arose. It is an excellent example of the intricacy of nature.

There are about a million different kinds or species of insects on earth. Because there are so many insects, a very elaborate system of nomenclature has been devised to categorize each species: the honeybee belongs to the Phylum *Arthropoda,* Class *Insecta,* Order *Hymenoptera,* Superfamily *Apoidea,* Family *Apidae,* and the Genus *Apis.* There are four species or kinds of *Apis.* All *Apis* are social bees with colonies not too much unlike that of our bee. Three of these species, *Apis dorsata, Apis florea* and *Apis indica* are Asian. The scientific name of our honeybee, the fourth species, is *Apis mellifera.*

The biology of the honeybee is discussed in greater detail in Chapter XIII.

The Value of the Honeybee to Man

The honeybee is not the only insect that can cross-pollinate flowers, but it is the most important. In fact, there are several thousand species of insects that carry pollen from one flower to another. Without honeybees and other insects to do this job we would not be able to produce a great variety of fruits and vegetables.

Of all the insects that visit flowers the honeybee is the only one that can be easily moved from one location to another by man. It is also the only insect species that man can husband and with which he can manipulate the population for his own benefit. In the western United States there are two species of wild, solitary bees that are used for the pollination of alfalfa and are believed to pollinate other crops. While these bees have certain advantages under the conditions in which they are used, one must still employ a good management scheme to make their use profitable.

One reason the honeybee is considered such an outstanding pollinating insect is that it shows flower fidelity. Several thousand years ago the Greek philosopher-scientist Aristotle observed that when a honeybee went to the field it would move from one clover flower to another clover flower and it would not, under normal circumstances, go from a clover flower to a flower of another plant species. Since pollen can fertilize an embryo only of the same species, it is important that good pollinating insects show flower fidelity. Most other insects that visit flowers do not do so. Certain insects that collect pollen to feed their young have been found to have as many as six or eight different kinds of pollen on their bodies after a trip to the field. This occurs only rarely in the case of the honeybee; the pollen on a honeybee's body would be all of the same type under normal circumstances.

The direction of agriculture in the United States is twofold. On one hand, we have fewer farms than were present before World War II, but they are much larger farms. On the other hand, we have an increase in the number of hobby or gentleman farmers. Honeybees will continue to play an important role in pollination for both. While some beekeepers will continue to make their living producing honey and beeswax, already we have many beekeepers who specialize in pollination and whose income is derived almost solely therefrom.

However, it is not only for practical agriculture that honeybees are necessary. The gentleman farmer with only one or a dozen apple trees must have bees for pollination as well as the grower with several hundred acres. At the same time, many of the wild fruits, nuts and berries that sustain our wild life throughout the year must also be insect-pollinated. The number of beetles, flies, wasps and solitary bees that pollinate flowers varies from one area to another. There is little that man can do to control the fluctuations in populations of these insects. This makes the honeybee especially important.

Who Keeps Bees?

It is estimated that about 300,000 people in the United States own one or more hives of bees; many own and operate one thousand or more colonies; there are a very few beekeepers in the United States with more than ten thousand hives. The total number of colonies is in the vicinity of 5.4 million and honey production is about 250 million pounds annually.

United States beekeepers produce about five million pounds of beeswax annually. Until 1971 beeswax sold for a price which was four to five times that of honey. We import about as much beeswax as we produce. We import honey also, and again, until 1971 the amount was seldom more than about 5 percent of our total production. In most years, until recently, the amount of honey exported was about the same as that which was imported. The world shortage of honey, which is responsible for the higher price of honey, first affected the market in the fall of 1971. Since that time statements about imports and exports, and the relative price of beeswax and honey, have had little meaning and have remained in a state of flux.

While California, followed by Minnesota and Florida, is usually the leading honey-producing state, no one area has a monopoly on honey production, number of colonies or the number of persons interested in beekeeping.

Only a few thousand men and women make their full-time living by keeping bees. Most keep bees part time because of their interest in nature and the pleasures of working with bees. Most also enjoy honey and the profits derived from renting their bees for pollination. The great majority of beekeepers are hobbyists with one or only a few colonies, who, despite varying backgrounds, share a common interest in one of the most interesting animals on earth.

Where May Bees Be Kept?

Two questions arise when we ask where bees may be kept. First, can a colony obtain sufficient food for itself and also enough surplus for the beekeeper to harvest; secondly, because many people fear bees, are there certain places where they should not be kept?

There are only a few places, such as deserts and heavily wooded areas, in the United States where a single colony cannot obtain enough pollen and nectar to sustain itself through the year. Mountainous areas would not have a sufficiently large foraging area for bees to provide a colony with much surplus honey but a single colony might survive under certain circumstances. Bees are even kept on rooftops in New York City; cities usually have a sufficient number of ornamental plants in parks and around private homes to sustain a few bee colonies.

Commercial beekeeping is undertaken only in locations where one may keep forty to fifty hives in a single location all year. For the commercial beekeeper success depends upon a sufficient supply of nectar. For example, alfalfa is one of our most important honey plants. Furthermore, alfalfa honey is one of the best honeys produced in the United States. Alfalfa has become an important dairy plant, especially on land that has a high lime content. Alfalfa should be cut for hay as, or just before, the flowers open; however, farmers are seldom ahead in their work and many alfalfa fields are cut much later than they should be from the point of view of high-quality dairy food. For this reason several million pounds of alfalfa honey are produced annually in this country. The beekeeper who is seeking a good location for his bees would do well to locate them on or near a large dairy farm where alfalfa is the major plant grown for hay. A good beekeeper must always be alert to the agriculture and land-management practices in his area.

Heavily wooded, mountainous areas and good farmland where there is a large acreage of alfalfa represent the extremes in poor and good areas in which to keep bees. If we examine the United States more closely, we find that each state and territory has certain localities where a special honey plant flourishes. For example, in the northeast the following are some of the important plants and areas: blue thistle (northern New York State), sumac (Connecticut and the Hudson Valley), purple loosestrife (along most large rivers and in swamps), buckwheat (western New York, Pennsylvania and Ohio), wild thyme (the western half of the Catskill Mountains, parts of western Massachusetts and east-central New York), basswood (scattered areas in the northeast), goldenrod (most of the northeast), clovers (most high-lime areas). Apple, willow, maple, dandelion, yellow rocket—all of which can produce a surplus of nectar—are generally considered buildup plants needed by the colonies in the early spring. County agents, conservation officers and state entomologists can pinpoint the particular areas where major honey plants grow in profusion.

The beekeeper should start with the premise that one, and usually two to ten colonies, may be kept almost anywhere. Above this number, only time and experience will show how much honey an area will produce. The beekeeper who wants a certain type of honey, or who wishes to produce large crops, may find it necessary to establish an out-apiary many miles from home. With a good knowledge of bee behavior one can successfully manage an apiary with a minimum of about eight visits per year; however, most beekeepers visit their locations sixteen or more times a year.

Many people are afraid of bees and the beekeeper must recognize this fact. The gift of a jar or two of honey each year will do much to alleviate a potential problem. A fence around an apiary will hide bees and forces them to fly above the heads of people in the vicinity. Bees require large quantities of water in summer and they can be a nuisance at swimming pools, birdbaths and small backyard fish ponds. Providing a place with fresh water in the apiary will eliminate neighborhood problems. A very few cities and villages have enacted ordinances against keeping bees within their boundaries; one such city is Washington, D.C. But then, no President since Franklin D. Roosevelt has been a beekeeping enthusiast. One may appeal a municipal decision against beekeeping, pointing out the value of honeybees as pollinators of fruit and vegetable crops and in pollinating the fruit, nut and berry plants used by wildlife. Still, the beekeeper has an obligation to keep his bees where they are not a nuisance, and this is easily arranged.

What Is a Good Apiary Site?

A good apiary site is secluded, exposed to full sunlight, has good air circulation and water drainage, a source of fresh water, and is in close proximity to a multitude of flowering plants. It is helpful if there is a small building nearby in which beekeeping equipment can be kept.

The apiary site should be secluded, because some people are afraid of bees, and others might vandalize the hives. While vandalism is not a serious problem, it is a temptation, especially for a young boy, to molest or even tip over a hive of bees. If any apiary site is hidden, this means that the individual bees that leave the hive must fly up and over surrounding vegetation. It means, too, that they cannot accidentally fly into someone walking in the vicinity. Herein we find an interesting aspect of bee biology. Close to the hive almost any bee is quick to defend the nest. A bee in the field is not so inclined. A bee disturbed in the field usually flees the site of danger or interruption as soon as possible. While it is true that an individual can be stung when not near a nest, it is a rare experience. Quite frequently, too, people who have been stung while walking through a field have offended a wasp, not a bee. Unfortunately most people do not know the difference between a wasp and a bee.

We have kept an apiary of twenty to forty colonies of bees only a short distance from the active part of the Cornell University campus for many years with no difficulty. Our apiary is surrounded by a hedge about fifteen feet high and about as thick. The hedge consists of evergreens, and inside the row of evergreens is a second hedge of deciduous bushes that grow to about ten feet in height. The bees must fly up and over the hedge to forage; they are also hidden from view. We have the room to grow such a large hedge, but in more confined areas a board fence would serve the same purpose. Commercial beekeepers often place their apiaries in a woods, usually close to a good road, but hidden just enough so the colonies cannot be seen by people driving by.

An apiary should be exposed to as much sunlight as possible. Foraging bees will fly to the field earlier in the morning and will work later in the evening if their hive is warmed by the sun's rays. This is especially true in the spring and fall, critical times for honeybee colonies. A sun-warmed colony with a large force of bees to send into the field will gather more honey than a colony that is shaded and cool and has a smaller field force.

Bees maintain a brood-rearing temperature of about 92 degrees Fahrenheit. If their hives are warm and dry, fewer bees are required to produce the energy to maintain this temperature, another reason sunlight is important. The maintenance of a uniform temperature within the colony is also important in helping the colony to control certain diseases that can occur when the brood-rearing temperature falls. Critical bee diseases —for example, sacbrood and European foulbrood—develop only in colonies under stress. Too cool a hive, because of an improper location, is one important stress that can be eliminated by the beekeeper.

Perhaps more important to the beginner is the question of hive temperament. Bees, or at least colonies of bees, have a temperament. On warm days, when the colony is able to maintain normal activities with little or no difficulty, the bees within the colony are much less inclined to sting. Experienced beekeepers will testify to the differences in stinging behavior between those exposed to sun and those in shade; bees in full sunlight always have a much better temperament.

Good air circulation and water drainage are important in an apiary. It is especially important to keep colonies of honeybees dry. Colonies that are damp or have wet bottom boards have difficulty in maintaining a normal brood-rearing temperature. A dry hive is a healthier hive. Honeybees also give off large quantities of metabolic water when they eat honey. It is important that water escapes from the hive and not condense inside. If moisture condenses in or near the brood-rearing area, it will cool the nest and make it more difficult for the bees to rear brood.

The best location for an apiary is on the side of a hill that slopes to the east or south and is devoid of trees in the immediate vicinity that might shade the location. While trees for a windbreak are helpful, they should not be too close to the site. Bees can be kept in less desirable locations but then the beekeeper must work harder.

Bees collect water to dilute the honey they feed to brood and also to air-condition their nest. In the spring, water may be a critical factor for a honeybee colony. If fresh water is not available nearby, it should be provided. In remote locations a fifty-five-gallon drum filled with water, and containing some straw, leaves or branches onto which the bees might crawl while collecting water, will provide the bees with water for up to a week or ten days. In the home apiary it may be possible to allow a water faucet hose to drip onto a long board from which the water may be collected by the bees. In the northern states beekeepers will notice that the number of bees at a watering site will increase greatly during July and August, when it becomes dry, indicating the need that bees have for water. The beekeeper who locates his apiary near a source of clean water will save his bees and himself much effort.

Without an abundance of pollen- and nectar-producing plants beekeeping is not possible anywhere. A colony of bees that produces a surplus of a hundred pounds of honey for the beekeeper probably collects five to six times that much for its own use. Honey is the chief food of the adult bee. Pollen is the bee's source of protein; it, too, is needed in quantity. Approximately one cell of honey and one cell of pollen are required to produce a young bee. While it is true that a bee may fly as much as eight or nine miles, if necessary, to collect pollen and nectar, research shows that colonies that gather most of their food within a radius of one half mile prosper much more than those whose field force must fly farther. Beekeeping is limited by the natural vegetation available to the field bees. Even the best of physical locations, without a sufficient number of foraging plants, is worthless. This is the key limiting consideration for the successful beekeeper.

Beekeeping Equipment

While men have been keeping bees for several thousand years, practical beekeeping has been possible only since 1851. In that year the Reverend L. L. Langstroth discovered the principle of bee space and invented the movable-frame hive. Prior to 1851 bees were kept in boxes or straw skeps; the bees attached their combs to the sides of the hive and it was not possible to inspect the interior of the hive or the brood nest. Because this was true, little was known about bee diseases and

Straw skeps have been popular in Europe for several centuries; they are rare today and are illegal in some American states because inspection of the brood nest and colony is not possible.

bee biology; the only way in which honey and wax could be harvested was to kill the bees or drive them from their nest. As a result, profits per hive were low prior to Langstroth's discovery.

Langstroth observed that there was a fixed space of about one-quarter to three-eighths of an inch between the combs in a natural nest. He noted that, if the spacing of the combs was not uniform, the bees would build a comb between two adjacent combs; such a comb is called a "burr comb" or "brace comb." Langstroth found that he could take a piece of natural comb from a nest and place it in a wooden frame; if this frame was placed in a box so that there was a bee space around it, and between the frame and the box, the bees respected and used this space. The bees do not build an additional comb when the bee space is correct. Thus our present hive has a series of movable frames and combs. Because of his discoveries, Langstroth has been called the Father of Beekeeping.

Today several firms in the United States make beehives. Many sizes and types of hives have been built for bees over the years. Today, however, over 95 percent of the beekeeping equipment made is the ten-frame Langstroth size. While this hive may not be perfect in every respect, bees adapt to it readily and it is a convenient size for a man to handle.

Beginners in beekeeping are advised to purchase only this standard size, the ten-frame hive. It has a greater resale value and parts are interchangable with new equipment. While it may be a temptation for a beekeeper to try to devise a more perfect hive, it is probably not possible to do so. When a beekeeper wishes to innovate, the place to do so is in his management system. No system of management is perfect; each region of the country is slightly different and a thorough knowledge of bee biology is required to manage a colony properly.

Three other important inventions, between 1851 and 1871, followed Langstroth's discovery: comb foundation, the honey extractor and the smoker. Of these, comb foundation is the most important. It was found that bees accept and build a perfect comb if they are provided with a thin sheet of wax onto which is embossed a six-sided cell base. These sheets of wax are called "foundation." While a colony of bees can build its own comb, it will not always build a straight comb. Also, there are two sizes of cells in an ordinary comb, worker cells (one-fifth of an inch in diameter) and drone cells (one-quarter of an inch in diameter). Since drones do not collect honey, the beekeeper should be interested only in producing worker bees. The best way to control cell size is to use full sheets of a well-made foundation.

Another important invention was the honey extractor. It takes a great deal of honey and energy for bees to secrete wax and build comb. When a honeycomb is used over and over— and some may remain in use for twenty to forty years—the bees are saved the effort of building new comb. An extractor is nothing but a centrifugal-force machine. If the wax cappings with which bees cover their full honey-storage cells are sliced off the comb, and the comb is placed in an extractor, the honey can be thrown out of the comb by centrifugal force. Thus the comb is saved and may be used many times.

The fourth and last invention important to the profitable and practical management of honeybees was the bee smoker. A smoker is nothing more than a firepot into which one may place punk wood, straw, hay, leaves, burlap or another material that produces a heavy smoke when burned. A bellows is connected to the firepot. This forces air through the fire and blows smoke out of a hole at one end. Prior to the invention of the smoker men used an open pot or held a piece of smoking wood or dung, blowing the smoke into the hive to calm the bees. While we do not fully understand why smoke calms bees, we do know there is no better method of doing so. Without smoke, at least on most days, it would not be possible to inspect a colony of

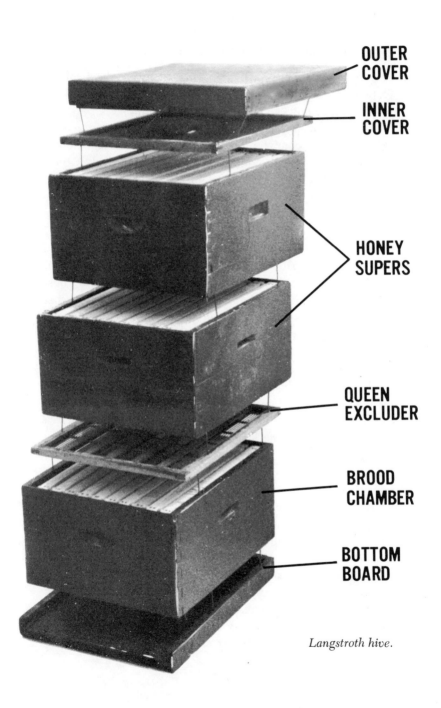

OUTER
COVER

INNER
COVER

HONEY
SUPERS

QUEEN
EXCLUDER

BROOD
CHAMBER

BOTTOM
BOARD

Langstroth hive.

bees. Many beginners have a tendency to oversmoke a colony. Only a small amount of smoke, a few puffs with most smokers, will keep a colony calm from one to several minutes.

These four major pieces of beekeeping equipment—the movable-frame hive, comb foundation, the extractor and the smoker —were all discovered within a relatively short period of time. Since that date there have been only a few minor inventions and discoveries in beekeeping equipment.

The important parts of the hive are the bottom board, the brood nest (box or super), honey-storage supers, the inner cover and the cover. A queen excluder—a series of·wires 0.163 to 0.167 inches apart, or a sheet of zinc with punched holes of similar width—is an important piece of equipment for most beekeepers. A queen excluder has spaces wide enough so that workers may pass through but queen and drones cannot. If the queen excluder is placed above the brood nest, the queen is confined in that area and cannot lay eggs in the honey-storage area. An excluder need be placed on a colony only three to four weeks before the honey is to be removed. Since the development of a worker bee takes twenty-one days, and a drone twenty-four days, any brood that is in the honey-storage area will emerge before the honey is removed from the hive. The use of an excluder is recommended, but not every experienced beekeeper will agree with this; some feel an excluder slows honey storage and cuts down on ventilation within the hive.

Each super of the hive will contain nine or ten frames or combs. While the standard hive was designed to hold ten frames—and ten should be used when they have new foundation—most beekeepers use only nine combs in the brood nest. The slightly wider spacing makes it easier to remove combs and to inspect the brood nest; yet it also means that the bees will build more unwanted burr and brace comb. However, there should be some rotation of combs in and out of the brood nest, making sure only perfect or almost perfect worker combs are present for egg laying. The rotation of combs with excess burr and brace comb into the honey-storage area each year—where they will later be uncapped and cleaned—lightens the chore of frame maintenance.

In the honey-storage area beekeepers use eight or nine combs, evenly spaced. By using an even wider spacing for combs in which the honey is stored, bees draw out the cells, making them deeper. It is easier to uncap thicker combs of honey.

Homemade Equipment

Many commercial and amateur beekeepers make their own beekeeping equipment. There is no reason why they should not do so. In addition to saving money, it is also possible to introduce innovations the beekeeper might find useful or helpful in his routine management.

There are two important considerations when making equipment. The first is to make sure that bee space is properly observed and built into all equipment. Second, and equally important, is that homemade equipment should be of standard size and interchangeable with all other equipment.

As with factory-made bee equipment, it is important that all new equipment be properly nailed. Equipment that comes into contact with the ground or dampness should be protected with wood preservative. The holding power of a nail lies in its length, not its diameter or the finish coat it receives; the longer the nail the better it will hold, and it is advisable in nailing any piece of beekeeping equipment to use nails as long as the equipment will take.

There are several wood preservatives that may be used to treat beekeeping equipment. Pentachlorophenol is the cheapest and best material to use; it is not toxic to honeybees. One must exercise care when using wood preservatives as certain of them also contain insecticides; this is a recent misfortune that has caused a good many beekeepers considerable grief.

Some of the pieces of equipment not usually available through the bee supply houses that can be made by the beekeeper include combination bottom boards and covers made with either shingles or quarter-inch plywood, special bee escapes, moving screens, special comb-honey supers, special bottom boards for comb-honey production, and inside furniture such as free-hanging frames. (A free-hanging frame has a top bar, bottom bar and end bars the same width, unlike a commercially made frame with wide shoulders.)

Hive Stands

The hive stand is important in many parts of the United States. Very few colonies of bees rear brood throughout the year; most colonies slacken their brood rearing during October, November and December. Whenever bees are rearing brood they attempt to maintain an interior hive temperature of about 92 degrees Fahrenheit. If the bottom board or super is damp, it is extremely difficult for the bees to do so; under damp conditions additional energy will be expended and the brood may be chilled.

A hive stand that keeps the colony of bees off the ground six to eight inches and tilts slightly to the front so that the water drains off the colony entrance with ease is recommended. Additionally grass growing in front of the colony entrance slows bee flight and a hive stand that raises the colony above at least some of the grass can help in this regard.

During the winter a properly constructed hive stand forms a dead air space immediately under two colonies pushed together for the winter pack (see wintering).

The hive stand recommended is 48 inches long and 20 inches wide; the stand should be at least 6 inches high. The 48-inch

Examining a colony of bees in two standard Langstroth supers. Note the examiner stands to the side of the hive with the smoker ready. Colony examinations such as this and without the use of a veil can be made with small colonies but should be done only by beekeepers.

hive stand is long enough so that the two colonies may be pulled apart during the summer, yet when pushed together during the winter a dead air space is created underneath if the two cross members of the hive stand are 32 inches apart center to center.

Since hive stands are on the ground and may rot quickly, they should be treated with either creosote, pentachlorophenol or some other good wood preservative. A hive stand properly treated will last for twenty or more years.

Maintenance of Equipment

There are several considerations regarding the routine maintenance of beekeeping equipment. While the physical appear-

A frame of capped brood (brood in the pupal stage); this is a very good brood pattern since only a few cells are vacant and the brood is compact. Note the queen cup in the middle left of the brood. The burr comb below the bottom bar indicates this frame was taken from the bottom of the hive.

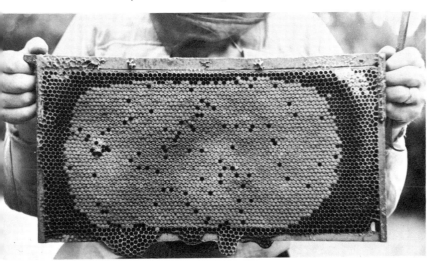

ance of the hives may be important to some, there is also the question of ease of management; additionally, the beekeeper should keep in mind the longevity and resale value of his equipment. At the same time, while honeybees may survive under even adverse conditions, it is advisable to take steps necessary to keep the hives dry; if the hive is dry, the bees will keep it warm without further assistance.

Routine maintenance of equipment does not refer merely to the painting and nailing of equipment, though these, too, are important. It is advisable to have on hand clean supers, bottom boards, inner covers and covers, which may be used whenever the equipment in use becomes too coated with propolis or wax, or needs painting or repair. As a routine practice it is helpful to change the supers in the brood-nest area every second or

Propolis and burr comb on a bottom board. The front two inches of this type of bottom board projects beyond the bottom super; thus the large amount of propolis on the left was deposited by bees to close partially the colony entrance. Such deposits are normal and will be rebuilt if removed.

third year. This is most easily done in the early spring when the colony population is at its lowest; it is only necessary to lift the frames from the old super and to place them in a clean one. So doing will make it easier to handle frames during the active season.

Spring is also an excellent time to check the quality of the combs in the brood nest. Combs with too much drone comb should be removed and replaced with good combs. European beekeepers often carry the replacement of brood combs to an extreme. In Europe, where generally speaking, honey crops are mediocre, or even worse, considerable emphasis is placed on growing worker bees as large as possible. Each time a worker bee is reared in a cell its pupal case is left in the cell. While each cell is cleaned before the queen lays in it again, the old pupal case is not removed. Over a period of years, and with a successive build-up of old pupal cases, the cells become smaller in size. This means that the worker bees produced in these cells will be smaller. Some European beekeepers routinely replace every comb in the brood nest after four years. In the United States, combs that have been used in the brood nest for twenty-five or more years are not uncommon. There is no doubt that old combs produce bees that are slightly smaller, but we do not feel this is an important consideration; the important thing in keeping bees is increased honey production, which depends on intelligent management. The routine maintenance of equipment should be for the purpose of making manipulations easier, thus aiding the beekeeper in his management scheme.

Wood Preservatives

Not only should hive stands be treated with wood preservatives but it is also advisable to treat bottom boards and wooden covers. Some beekeepers treat their supers, or at least the corners of the supers. If pentachlorophenol is used as a preservative, it can be purchased as a concentrate to be diluted with kerosene. If the bottom boards are cold-soaked for twenty-four hours, they will have a life of two to three decades. After the soaking process the bottom boards should be allowed to dry and weather outdoors for three to four months; it takes a long time for kerosene to evaporate.

Creosote is also a good wood preservative, but it has the disadvantage of making the equipment dirty to handle. There are several other wood preservatives available but they are more costly.

Painting

Well-painted beehives will have a longer life. If different colored paints are used, the bees will be aided in orientation, and drifting between colonies will be reduced. Usually bee-keepers paint the top and bottom rims and outsides of supers; the inside is usually left unpainted. Placing some paint in the rabbets or frame rests will usually deter the bees from depositing too much propolis at that point for a year or two.

Standard Dimensions

Over the years beekeepers have patented more beehives than any other piece of apiary equipment. Many people have attempted to create the perfect beehive, which would be most acceptable to the bees and from which the greatest crop could be secured. However, it is now generally agreed that management of colonies is far more important than the type of equipment in which the bees are kept.

Recently manufacturers and state apiculturists have sought to standardize the equipment beekeepers use. While all manufacturers do not make uniform equipment, there are generally agreed-upon dimensions.

Dressing for the Apiary

Honeybee temperament is affected by temperature, humidity, colony prosperity, pests that irritate the colony, the availability of food and other such factors.

Generally speaking, it is best to inspect a bee colony near midday on a day when the sun is shining and when there is good flight to and from the colony. It is usually the older bees in a colony that protect the nest and are, therefore, more apt to sting. If most of them are foraging, the beekeeper will have an easier time. However, if there is a dearth of nectar, even an active colony is difficult to manage on a warm day. If nectar is being brought into the colony in quantity by field bees, there are very few guard bees and very little problem with stinging bees.

It is well known that honeybees, as well as other stinging insects, are less prone to sting light-colored, smooth-finished materials. Khaki clothing is a favorite with beekeepers for this reason; white coveralls are also popular. At the same time, rough materials such as leather, suede, felt, etc., appear to irritate bees and they will sting these materials more readily.

The two extremes of the body, the ankles and the face, appear to attract bees. It has been suggested that the blinking of the eyes and the protrusions of nose and ears are the reasons people are often stung in the face. A pair of boots and a good veil will protect against stings in sensitive areas. Veils are made of black cloth or black wire over the face since it is difficult to see through any other colored material. A few beekeepers wear gloves but they are not advised. Gloves are cumbersome; of course, one will receive a few more stings if gloves are not worn, but the stings are infrequent and not of consequence. A gloveless beekeeper should be more aware of the colony's temperament and will use smoke as it is needed.

Bee Stings

When an individual honeybee or a honeybee nest is attacked, the bees have two choices: they may attack, or they may flee. Interestingly enough, honeybees may flee as a group if they are disturbed too much; an individual bee disturbed in the field will almost always abscond if given an opportunity. Since the sting is the honeybee's major method of defending both itself and its nest (even though it results in its death), it is not unreasonable that a beekeeper should expect to be stung occasionally.

Commercial beekeepers or bee inspectors, men who work with bees every day, are stung frequently, often more than a hundred times a day. However, they soon build an immunity to swelling from bee stings, and the experienced beekeeper cannot tell where he was stung a minute after he has been hit by an angry bee. Being stung is part of keeping bees. The number of stings one receives may be cut to a minimum, but much less than ten or twenty stings a year is insufficient to build up an immunity to the swelling and pain associated with stings.

It is not impossible—though perhaps it is improbable—for a man with one or two colonies to work through an entire season without being stung. To do so would be to miss part of the action; while a good beekeeper does his best to avoid excessive stinging, he does not try to avoid the occasional sting. For some, it is part of the initiation into the beekeeping fraternity.

Clothing, use of smoke, time of day, the condition of the colony and the skill of the beekeeper are all considerations in making colony manipulations. Many stings can be avoided, but to avoid all stings would be to expect too much.

There is little one can do to prevent swelling, pain or the itching from a sting if one has not developed an immunity. A sting, if visible, should be removed. Ice applied to the area stung may slow the distribution of the venom in the system; however, the chief purpose of applying ice, mud, a cold pack or hand pack to a sting is to create a new sensation. This can be very helpful.

Outbuildings

Prior to the time that beekeepers used trucks to visit their out-apiaries, the construction of a small building in each out-yard was practiced by commercial beekeepers. Usually extracting was done in these locations and the building was equipped with necessary tools and extra hive equipment. Even today beekeepers find that buildings in out-apiaries are convenient. One usually needs a place to keep extra bottom boards, covers, inner covers, supers, etc. so they may be available when needed. Also, if a hammer and saw are present, repair of equipment may be made when needed. Perhaps most important, the beekeeper may store his supers of combs in an out-apiary. While one may purchase insurance against the loss of drawn combs, insurance does not replace combs as rapidly as they are needed when colonies build up in the spring. A beekeeper who stores all of his combs in one location and loses them because of fire or theft may be forced out of the business. Thus commercial beekeepers generally divide their supplies among several small buildings.

The major problem with outbuildings is that they may be vandalized by wildlife and man. If the building is in a permanent location, a cement floor is advisable. When windows are installed, heavy shutters should be put over the outside so that vandals cannot enter the building, at least with ease. In certain parts of the country raccoons and squirrels have been known to make their nests in outbuildings and for this reason it is important that the buildings are secure.

II

How to Start in Beekeeping

There are basically four ways in which a person interested in beekeeping can obtain new colonies: he can buy a package of bees that contains a queen and several thousand worker bees from one of the southern states; capture a swarm of bees; remove a colony of bees from a bee tree or building; or, and perhaps best, buy a secondhand colony of bees.

There are several precautions to be taken in each case, but perhaps most important is the question of obtaining or buying disease-free equipment or bees. While honeybees have many diseases, there is only one we really fear—American foulbrood. This is a bacterial disease and is widespread in the United States.

States from which package bees are shipped have apiary inspectors who certify that the bees are free of disease; the accuracy or correctness of such a certificate will vary from one state to another, but generally certificates of disease infection are valid. Whether or not a bee tree, or bees in a building, or bees in a swarm harbor disease is always a good question; such bees should be watched carefully for the first several months after they are removed from their home in the wild. Most states require that bee colonies be inspected by a state apiary inspector before they are sold or moved; when buying bees it is always best to write to the apiary inspector at the state capital to determine state policy in this regard.

From the point of view of time and money, probably the best way to start in beekeeping or to obtain more bees is to buy secondhand equipment. Such equipment will sell for one-quarter to one-half the cost of buying a package of bees and new equipment. While the bees in a captured swarm or bees from a bee tree may cost nothing, there is the cost of the equipment; this, of course, can be secondhand and is usually inexpensively purchased from a local beekeeper.

An added advantage of buying secondhand equipment is that the beginner can seek advice and sometimes help from a more experienced beekeeper. This is especially important when the colonies are moved to their new location and when the first inspection is made.

The Best Time of Year to Start Beekeeping

The best time of the year to buy a hive of bees, install a package of bees or capture a swarm is in the early spring. This means April in most of the northern states, earlier in the southern states. The reason for recommending this time of year is simply that the major honey flow begins in late June or July. Instituting a reasonable management scheme early in the spring enables the bees to take maximum advantage of the available nectars.

While there is a recommended time to buy an established hive of bees, this again is only because of preparing the colony for the honey flow. Secondhand colonies of bees may be purchased any time of the year, but brood should be present to allow inspection for disease. The price of an old colony may be less in the fall, but one must also contend with the problems of wintering and the danger of winter loss.

One should not expect a swarm or a package of bees hived

after the first of August to survive the winter without extra attention. Swarms and packages established in May and June should be able to collect sufficient food to sustain themselves through the winter, but they will seldom gather a surplus of honey the first year.

Buying a Hive of Bees

Buying a hive of bees is the best and cheapest way to start beekeeping or to increase the number of colonies a beekeeper owns, provided, of course, that the price is right and the equipment is free of disease.

Most colonies are sold in two boxes or supers; the price for

Two wire-bound packages of bees, each containing about three pounds of bees and a newly mated queen. Packages are produced in the southern states in the spring for use in the north to start new colonies.

LEFT: *After the small cage with the queen and the feeder pail are removed from the package, the bees may be shaken into a colony.*
RIGHT: *Two tinned pails filled with sugar syrup. Many holes in the lid of the pail allow the bees to suck the syrup from the pails which are placed on top of frames in the hive with the newly installed package of bees. The package, with the few remaining bees inside, is placed in front of the colony so the bees may enter the colony.*

such a unit is seldom under ten dollars and rarely over 25 dollars. In addition to having the bees inspected by a state apiary inspector, the beekeeper should check the colony to satisfy himself that it has a queen and that sufficient food is present. Ten to fifteen pounds of honey should be present at any time of the year; in the fall a colony needs sixty pounds of honey to survive the winter.

Since over 90 percent of the beekeeping equipment used in the United States is the ten-frame Langstroth size, it is advisable to purchase only this size hive. If the supers are not a standard size, the price should be discounted. Additionally, the price should be determined by the physical condition of the supers and also the condition of the combs within the colony. Colonies with old combs, broken combs, combs chewed by mice or with too much drone comb should be replaced, and replacement can be expensive.

The ease with which the colonies can be moved and any assistance the beekeeper gives in moving the hives are again factors to be considered. While buying secondhand equipment is the best way to start beekeeping, it can also be the most difficult and perhaps most expensive for the uninitiated.

Buying and Installing Package Bees

Package bees, bees sold by the pound in wire-covered boxes, together with a queen, are available from a number of general merchandising houses, bee supply companies and independent beekeepers in the southern states. While packages are occasionally sold by a beekeeper in one of the northern states, most of the package-bee-producing states, with the exception of California and Georgia, border the Gulf of Mexico.

Packages are sold in units that contain from one to five pounds of bees; there are about four thousand bees in a pound. Generally, for the beginner, a three-pound package with queen is most satisfactory. Packages are shipped by mail or rail freight; they are usually two to three days in transit and do not appear to suffer from confinement for this period of time. The package contains a small can of sugar syrup the bees use as feed. Bees in transit in packages usually do not need water. The queen is held separately in a queen cage within the package; the queen cage contains sugar candy on which the five or six worker bees in the queen cage may feed.

As soon as the package of bees is delivered, it should be placed in a darkened room with an ideal temperature of 70 degrees Fahrenheit. The bees should be placed in their new hive on the day they arrive. They should first be fed by brushing sugar syrup, a mixture of about half water and half sugar (dissolved in hot water), over the wire surface. Usually a three-pound package of bees will consume a pint or more of feed which should be given to them in four or five feedings over a period of an hour or more. The bees in the package will be much more docile and easy to handle when placed in a hive if they are well fed.

The bees in the package should be placed into the new hive in the late afternoon, about half an hour to an hour before dusk. The reason for this is that in a strange location the bees that take wing without orientation can become lost or drift to another hive. If more than one package is being installed at the same time, the hives should be several feet apart and/or should be painted different colors. In the event of cool or rainy weather it is better to go ahead with the installation of the package than it is to delay the matter. It is most important that the bees begin brood rearing as soon as possible so that the replacement of the old and dead bees takes place without delay.

Package bees should probably be put on frames with new foundation, not old comb. In this way, if the package bees are carrying any honey with disease spores, the honey will be used in the construction of new wax and the disease spores will be lost. If package bees are installed on old combs, there is always the possibility of the bees' depositing any honey they carry in the cells; such honey, if it contains disease spores, can later infect the colony. Obviously, installing package bees on foundation delays the development of the colony; brood rearing starts sooner if the queen has old combs in which to lay eggs. For this reason many beekeepers install their packages on old combs, watch the developing packages for disease, and hope they did not buy bees infected with American foulbrood. The quality of bee disease inspection differs from one state to another and so it is better to buy bees from a known source rather than through a common supply house.

Some beekeepers use only four or five frames in a hive when installing a package, while others put a full complement of combs in the hive to start; whichever is done is probably of little consequence. If only four or five frames are used at first, others must be added as needed. The actual steps in the installation of a package are as follows: feed the package, remove the wooden cover over the feeder pail; remove the feeder pail, shaking any bees that cling to it into the ready hive; remove the queen cage and make certain the queen is alive; remove about one-half to two-thirds of the candy from the queen cage so the queen will be released by the bees within about twelve hours; place the queen cage candy end up between two frames so that the screen face of the cage is exposed to the bees; shake the remaining bees from the package into the hive. It may be difficult to shake the last hundred or two hundred bees from the package. If the empty package is placed in front of the hive with the hole in the cage facing the hive, the bees will crawl into the hive if the weather is not too cold.

No smoke is needed to install a package of bees. The bees should be well fed as indicated; well-fed bees are gentle bees and cause no difficulty.

After the bees have been shaken into their home, one or

preferably two pails of sugar syrup should be placed on top of the frames and inside an empty super placed on top of the new hive. A newly installed three-pound package of bees will consume two ten-pound pails of sugar syrup within a week or ten days and should be fed again when the pails are emptied the first time. Whether or not a third feeding is required depends upon how much natural nectar is available. In some areas packages may need a third feeding and may consume twenty to thirty pounds of sugar in total. It may be helpful to place a burlap bag over the feeder pail and the tops of the frame. The entrance of the colony should be reduced to an opening of three-eighths by three inches. The colony should not be inspected for at least ten days, although the feeder pails might be checked after five to seven days.

The purpose of the first inspection of a package of bees is merely to determine if the queen is alive and laying. Unfortunately, the success of the package rests entirely with the queen. If the queen fails, the package is lost. If, upon the first inspection, the queen is not present and laying, the only alternative is to combine the package with another package or colony. There is no other alternative (see uniting colonies).

Capturing a Swarm

Hiving a swarm of bees can be a thrilling experience. It is not difficult and requires only a little knowledge of beekeeping. It is not uncommon to find a swarm of bees hanging from a limb in a tree or bush in the late spring and early summer months. Once found, it is usually possible to entice the bees to accept a man-made home.

Swarming is the natural method of colony reproduction in honeybees. While it is important that young bees be produced within the hive, before man came on the scene it was also im-

Comb built in a hollow tree by bees that have probably swarmed from a beekeeper's hive. While the comb in the nest is somewhat irregular, there is still a bee space between the combs. The bees in this nest had starved because of a lack of food in the early spring.

portant for colonies of bees to swarm; otherwise, the species could not perpetuate itself. There must be both reproduction within the colony and reproduction of the colony.

While it is easy to find a swarm in the vicinity of a commercial apiary, one should not be surprised to find a swarm almost anywhere. One of the favorite nesting locations for escaped bees is in the sides of a house, especially an old house where there is space between the studs. Thus, many swarms are found in cities. Once a swarm is found, one cannot delay too long in attempting to capture it. A swarm will move into a new home, an abandoned hive, hollow tree or the side of a house, but it will usually require two to three days to do so.

Bees in a swarm gorge themselves with honey before they

Natural comb built by a swarm in a hive without frames. Note the circular shape of the comb and the fact that even in nature the combs are parallel. Colonies may survive in an old hive in this manner, but it is illegal in most states to keep bees in such a hive since the combs cannot be inspected for disease.

leave the parent hive; bees engorged with honey are gentle bees. However, if the bees have been away from their hive for some time and because of inclement weather have been unable to gather food, they are very prone to sting. Severe stingings by honeybees are not reported too often; however, those that do occur are usually the result of someone's attempting to hive a swarm which has exhausted its food supply. Beekeepers call swarms without food dry swarms.

One can determine whether or not a swarm is with or without food by being aware of the weather conditions for the day previous to the time one tries to capture the swarm. Bees in a swarm forage; they share their collected food with the other bees in the swarm. If the bees in a swarm have been able to

Natural comb built by a swarm in a tree. Note the bee space between the combs. Bees rarely build comb in an unprotected area where they cannot survive the winter; they do so only when the light level is reduced at the time they initiate comb building.

forage, they will have food and one can capture and place them in a hive without difficulty.

Hiving a swarm is not unlike hiving a package of bees. The bees in a swarm can be placed on foundation, thus partially eliminating the danger of disease begin carried into the new hive, or the bees can be hived on old combs. Since swarms are usually hived later in the year, it may or may not be necessary to feed the bees in a swarm. Only the experienced beekeeper can tell whether or not food is needed.

To hive a swarm it is only necessary to shake the bees from the branch upon which they are found into a new hive. If the super used for the new hive is an old one, a super that has a bee odor, or if there are old combs within the hive that have a bee odor, the bees will accept their new home much more rapidly. Odor is important in the social organization of the honeybee colony and bees prefer to nest where bees have nested before.

If the swarm of bees is on a limb close to the ground, it is only necessary to bend the limb and shake the bees with one or two hard shakes into the hive after the cover has been removed. If the swarm is higher in the air, one has the alternatives of taking the new hive or home up a ladder, or of cutting off the limb on which the bees are hanging and carrying it to the hive. If the limb can be cut with the bees intact as a group, then the bees may be shaken into their new home without difficulty.

Not infrequently bees that have been shaken into a hive will not accept the new home; they will leave the hive and the cluster may reform in the old location or a new one. Often the second time a swarm is shaken, the bees accept the new home. As with hiving packages, the beekeeper will have greater success hiving a swarm in late afternoon.

When hiving a swarm, it is advisable to staple the bottom board to the hive. If this has been done, it is a simple task to pick up the newly hived swarm at night and to move it to the apiary. It is usually not necessary to screen the entrance or to place a top screen on the hive (see moving bees), but this can be done if the beekeeper wishes.

Bait Hives

Bees prefer to nest where bees have nested before, presumably because the odor in an old hive or nest is attractive to them. Thus, if a colony of bees in a tree or the side of a house dies, one can reasonably expect a new swarm to take up residence in the old location within a year or two. Beekeepers who store their empty combs outdoors often find a stray swarm has moved into them.

A bait hive—an old hive containing a single comb—is attractive to a swarm. Some beekeepers use bait hives, scattering them around the countryside to pick up stray swarms. It is not advisable to use a good super or good combs for bait hives, as they are also attractive to moths, mice, birds and other insects. However, old combs and old hives can be used in this way.

While every beekeeper tries to prevent swarming, he is not always successful. It sometimes pays to keep a bait hive in an apiary just in case an error is made. Since bait hives are made of old equipment, they are not satisfactory for a permanent home for a colony; however, the bees can be easily transferred to a new hive by placing a super of combs on top. The bees will soon move upward and then the old bait hive can be discarded.

Removing Bees from a Bee Tree or Box with Fixed Combs

It is possible to salvage the bees in a bee tree or box with fixed combs by one of two methods. Considerable time and effort are required to do so, but in both cases there is much to observe and learn about bee behavior. In the case of a bee tree it is necessary to cut down the tree and to cut the log so as to expose the top of the nest. The portion of the tree containing the swarm is then set upright again. Sometimes felling a tree to recover that portion containing the swarm so damages the comb that many bees are killed; this depends upon the size of the tree and how it is felled. When attempting to save the bees from a nest in a box or field crate, it is again necessary to remove some of the top boards so as to expose the combs.

When the combs in the top of a nest in a tree or box are exposed, one can drum the bees out of their old nest, a method that can take ten to twenty minutes, or one can place a super of old combs on top of the hive and let the bees move upward by themselves; the second method of transferring can take several weeks.

Drumming is a method of moving bees from fixed comb hives that is several centuries old. If one beats rhythmically on the sides of a hive with his hands or with a hammer, the bees will soon walk upward. One may place the new hive or home, or a cardboard box above the old nest and the bees will move into it. The beat need not be too fast, probably forty to sixty beats a minute is satisfactory. The queen will usually leave the hive with the rest of the bees. When bees are drummed, very few of them take wing; drumming does not anger the bees. Often about 10 percent of the bees, the younger bees, may not leave the hive.

If a hive is drummed, one may or may not save the old comb. It is possible to cut the comb, especially the comb containing brood, from the old nest and to tie it into a frame with string. The bees will later fix the pieces of comb into place. However, such combs are seldom perfect and the beekeeper often ends up with combs with too many drone cells. It is probably best to try to save only the bees and to render out the beeswax from an old nest.

A second method of removing the bees from a fixed comb nest is to allow the bees to move upward, by themselves, into a super of combs that has been placed above the old nest. It is actually only important that the queen move upward and begin to lay eggs in the new super; if she does not do so within a week, it is usually because there is a barrier between the old nest and the new, a barrier of too much honey, too little space or some other type of constriction. If eggs are not found in the new super within a reasonable period of time, some effort must be made to encourage the bees upward.

Once the bees have moved into the new super, a queen excluder should be placed between the two units; the queen may or may not be found before this is done. If she is not found, the beekeeper should check the hive again after three days to determine if there are eggs in the new hive. If eggs are present, the queen is present; if there are no eggs, the operation must be repeated until the queen is found above her old nest.

It is then necessary only to allow the brood to hatch out from the old comb below and then to destroy the old nest. Allowing the bees to move upward by themselves has the added advantage of saving the brood and probably results in a stronger colony of bees. As in the case of buying bees, placing bees from a fixed comb nest into a new hive is best done in the spring so that the bees have time to adjust to their new home.

Races, Varieties and Strains of Bees

The honeybee kept in the United States evolved in the Near East. It spread from there throughout Europe and Africa. When North America was settled, bees were brought here and they were also carried to South America, Australia and New Zealand. As a result of isolation of some bees many thousands of years ago in Europe, certain groups began to show differences. People have examined these differences and as a result have named certain races, strains or varieties, depending upon what they thought was the proper terminology.

Differences in bees do exist. Some bees, notably the Caucasians, collect and use far more propolis. Some races have a greater tendency to abscond or to swarm than others. Certain races of bees stop brood rearing when nectar ceases to be available while others use all available stored honey to continue to rear brood. The Cyprian bees and the African races are noted for their bad temper.

What bee should a beekeeper buy? The author takes the position that management is far more important than which queen is purchased. Secondly, the conditions under which a queen is reared are more important than the race. We know that queens that are properly fed during their larval stage will be bigger and able to lay more eggs than queens that are not properly fed. Finally, the author dislikes races of bees that use large quantities of propolis. Propolis makes it difficult to manipulate the combs and slows work in the apiary.

During the past eighty or ninety years the Italian bee has been the one most widely used in the United States. One of the major reasons for this is that Italians are fairly resistant to European foulbrood, a stress disease that can cause difficulty under adverse conditions. Furthermore, Italian bees are fairly gentle, though not the gentlest of bees, and they are good collectors of nectar. They appear to winter well in the north and are probably the bees to be recommended. However, there is no unanimity of opinion in this regard and beekeepers will no doubt continue to experiment with other bees as they become available to them.

III

Spring Management

Management is the key to successful honey production. Men have spent much time inventing new types of hives and selecting special strains of bees for specific purposes. While these efforts have not been wholly in vain, they are of less consequence than good management. Colonies of honeybees do not need much management; in fact, it is easy to overmanipulate a hive. However, at the right time, providing room, reversing supers and certain other steps are critical.

A beekeeper's year is said to begin in August for it is at that time of year that colonies should be requeened and, in effect, prepared for the following spring. Since most bees are bought and sold in the spring as either old colonies or packages, it is best to begin a discussion of management by first discussing the spring months.

The First Spring Inspection

The purpose of the first early spring inspection is to check for food, to pick up dead colonies and to combine weak colonies. While colony entrances can be checked occasionally throughout the winter, it is not until April, in most northern states, that it is possible to make sure that colonies have wintered satisfactorily and have sufficient food. Most beekeepers prefer to delay opening their colonies until mid- or late April, usually about two weeks after the first spring pollen is available. In fact, it is a reasonably correct assumption that if bees from a colony are gathering pollen, the colony is in good condition, except for the possible danger of being short of food.

The first inspection of a colony should be brief, especially if the temperature is less than 80 degrees Fahrenheit, an unusually high temperature for mid-April. The colonies should be unpacked, bottom boards and inner covers checked to make sure they are dry, and if not, they should be replaced. The colony should be hefted to determine if it has sufficient food; in mid-April a colony should have a minimum of 20 pounds of honey and hopefully more. It is probably not advisable to inspect the brood nest so early unless there is some special reason for doing so. Chilling of the brood can occur and cause much loss to the colony if the inspection is too lengthy; it is not until May that the equipment and colony condition should be checked more carefully.

The entrance cleat should remain in place for at least a month even after the colony is unpacked. The size of the entrance needed will depend upon the population of the colony.

Dead colonies should be carefully inspected to determine why they died; depending upon the area, American foulbrood may be the cause of the colonies' dying in the winter. Weak colonies should be combined, either with strong colonies or other weak colonies; they can be split at a later date and be given new queens. Trying to carry a weak colony through the spring buildup period is not worth the effort and is often futile. Colonies can be combined by setting one on top of another with only a sheet of newspaper between the two brood nests; three or four slits should be made in the paper. The queens will eventually fight and usually the stronger of the two survives.

Spring Feeding

It is not known how much honey a colony of honeybees actually needs to sustain itself during the year; however, the consensus is that a colony that produces one hundred pounds of surplus honey for the beekeeper probably collects many times that amount for its own use during the year. A general recommendation is that a colony of honeybees should have fifteen to twenty pounds of honey in reserve at all times of the year.

In many parts of the United States colonies can suffer from a lack of food in the late spring. This is especially true in the northern United States, where not infrequently there is a dearth of pollen and nectar near the middle or end of May and following the dandelion and yellow rocket bloom. Sometimes it is as long as a month after dandelion and yellow rocket bloom before the clovers begin to flower and yield nectar. While there may be some plants in flower during this period, there are relatively few. When a colony of honeybees goes into a starvation condition, their first reaction is to remove the larvae and discard them in front of the hive. The queen continues to lay eggs but as they hatch, these larvae too, are discarded. Eventually, of course, adults starve, though they can subsist for a certain amount of time on body reserves. The overall effect of this spring starvation is the weakening of the colony, perhaps even its death. Because bees rear large quantities of brood during May and June, they also consume large quantities of honey and it is important that the beekeeper be alert as to the food reserves in his colonies.

Many beekeepers save combs of honey from the fall for spring feeding. This is probably the easiest way to feed bees since only one inspection of the colony is necessary, at which time one or more combs can be inserted. An added advantage of feeding honey in the comb is the fact that old capped honey will usually not stimulate robbing. If sugar syrup is dripped on the ground in the apiary in the process of feeding, it is possible that robbing may be stimulated, especially if there are very few flowers in bloom. When honey is not available for spring feeding, then sugar syrup should be used. In the fall when bees are prepared for winter and sugar syrup is fed, the mixture is made up of two parts sugar and one part water; for spring feeding the ratio should be one part sugar and one part water.

There are several ways to feed sugar syrup to bees. Probably the two methods most widely used are the division board feeder and feeder pails that are put over the top of the colony. Both appear to work about equally well. Beekeepers who routinely feed in the spring sometimes leave their division

A division board feeder, which may be used to feed packages or colonies in the spring or fall. It fits into a hive as a frame does, and is then filled with sugar water.

board feeders in the colony all year. If pails are used for feeding, they should hold at least ten pounds of sugar syrup.

Another method of feeding sugar syrup is to pour it directly into the combs and to insert combs filled with sugar syrup along the edge of the brood nest or above or below the brood nest. Sugar syrup can be poured into the combs in the home apiary or honey house and then carried to the field with little loss. It is important when pouring syrup into a comb to use a sprinkler can or pail with many holes in it; a solid stream of syrup poured onto a comb will not penetrate the cells. Sugar syrup should be poured into both sides of the comb and then the comb should be given a good shake over a water tub or similar receptacle to catch the surplus syrup. If this is done, only a very small quantity of the sugar syrup will drip out of the comb while it is in transit to the apiary.

Some beekeepers recommend the feeding of dry sugar. It is true that honeybees do eat dry sugar and they will do so when there is a dearth of nectar and when they are on the verge of starvation. A few pounds of dry sugar can prevent starvation. And dry sugar should be used if absolutely nothing else is available. However, it is also correct that bees will often carry dry sugar out of the hive and discard it in the front of the hive. If other sources of nectar become available, bees will usually cease to collect or store the dry sugar that has been placed on the inner cover or bottom board of the colony. Bees must also collect water to liquefy the sugar in order to use it for food and this places a small strain on the colony.

Small entrance feeders for colonies, such as Boardman feeders, are not recommended except in emergencies. Even then they are usually useless. Colonies that need feeding usually require far more food than can be fed through a Boardman or entrance feeder. Furthermore, entrance feeders are cold at night and the bees will cease feeding from them shortly before sunset and will not feed again until there is warm weather.

They will not feed from them at all during inclement weather. Entrance feeders give the beekeeper an opportunity to observe how much food his bees are taking; they are used by a few beekeepers in the warm, southern states but are generally of little value.

Feeding Pollen Substitutes and Supplements

Pollen provides bees with the protein they need. In the beekeeping literature there are references to both pollen substitutes and pollen supplements; there may be some confusion over these two terms. A pollen substitute is a material used in place of pollen whereas a pollen supplement is usually the same materials fed to bees but with previously bee-collected pollen added.

The feeding of pollen supplements and substitutes is recommended more in some areas of the country than in others. In many states one of the major deterrents to beekeeping is the shortage of pollen during certain times of the year; Arizona is an example of a state where pollen supplements are fed in quantity. In other states bees may collect so much pollen as to become pollen-bound (to have too much pollen) and on occasion beekeepers have been known to discard combs of pollen; this has been known to occur only rarely. In the northern United States bees usually collect pollen in sufficient quantity throughout the active season that it is not necessary to feed pollen substitutes or supplements. In certain of the midwestern states these materials are recommended, but this is largely because some beekeepers in those areas use two-queen colonies and try to build larger populations of bees, especially in the early spring.

Generally speaking, the use of pollen supplements and substitutes is not advised for beekeepers in the northern United States; but opinions on this matter vary widely. The major reason for not feeding pollen supplements and substitutes is that it is not usually necessary. Also, there is the danger of building too large colonies, which may suffer from starvation in later May and early June.

Research on pollen supplements and substitutes indicates that soybean flour produced by the expeller process is satisfactory as bee food. The material is not too attractive to bees and for this reason adding bee-collected pollen to it is advisable. There are various methods of storing bee-collected pollen, including freezing, drying and mixing with sugar; probably bee-collected pollen that is stored in a freezer keeps best.

Adding honey to a pollen supplement will make it more attractive to bees; however, there is a real danger in using honey instead of sugar syrup since it is possible to transmit disease in this way. If a beekeeper is certain that he has no disease in his apiary, then honey might be helpful.

People in many countries are working on this question and research on pollen substitutes and supplements can make some major changes in beekeeping in the next few years.

Clean Equipment and Good Combs

Replacing worn and warped equipment, broken combs, combs with too much drone comb, scraping and cleaning hive bodies, inner covers and other equipment are jobs most easily done in the spring. At that time of the year there are fewer bees in the hive and changes can be made with the least difficulty. Keeping equipment in good condition facilitates work in the apiary during the active season.

During the course of the year bees collect and bring much propolis into the hive. An accumulation of propolis, together with an excess of burr comb, can slow manipulations and the removal of combs from a hive. Protruding pieces of burr comb can also kill or maim some bees as frames are removed from a super for inspection.

While it is normal for a colony of bees to contain some drones, the drone population should be kept to a minimum, as drones serve no useful function other than to mate with virgin queens. The easiest way to control the drone population is to have good frames with a maximum of worker cells in the brood nest. The best frames are made when full sheets of foundation are used. (See "Adding Foundation and Making New Combs," page 72.)

While replacing worn equipment and keeping good combs may appear to be minor aspects of spring management, it is part of the routine of careful management that leads to the production of a maximum crop.

Swarming

Preventing swarming is the most difficult task beekeepers face. This is especially true in the northern states and Canada, as swarming becomes more intense as one moves north. In Florida and the other southern states, swarming can occur during any of the spring months, but it is not the problem it is farther north. In part this is because colonies in the north usually contain more bees and are more crowded at the peak of the season.

Swarm control is an important part of spring management and often the most frustrating. In the early spring the beekeeper's main concern is that his colonies survive the winter and start to increase their populations in preparation for the first honey flow. Young queens, food, equipment, disease control and locations are all important in this regard.

However, once the colony is over this early building period, the beekeeper finds himself faced with the other extreme—

colonies that are too populous and are about to swarm. Swarming is a natural process. Without swarming, or colony division, the species cannot survive. Reproduction of individuals in the colony is not sufficient; there must also be reproduction of colonies. Only populous colonies will swarm under normal circumstances. Therefore, the beekeeper who builds colonies with large numbers of bees, which is necessary to secure the largest crop, is also creating conditions conducive to swarming. This is a simple fact the beekeeper must accept and around which he should build his management program.

From the practical point of view, swarming should be prevented. However, what is done depends upon the condition of the colonies and whether or not signs of swarm preparation are evident. Not all colonies of bees will swarm even though outwardly conditions are similar. It is in such cases that the beekeeper's skill in diagnosing what is taking place is important.

Determining whether or not a colony will swarm is done by observing the construction of queen cups and queen cells in or adjacent to the brood-rearing area. Where colonies are kept in two or more standard Langstroth supers, following the development of cups and cells is most easily done by tilting the upper super upward and forward and looking at the bottom bars of the exposed frames. Since there is space between the supers, and the area in the middle of the hive is still in the brood nest, this is the most likely place where cups and cells will be built.

Swarm Control Versus Swarm Prevention

Swarm prevention is concerned with those steps taken to deter or prevent queen-cell construction within a colony; swarm control involves the steps the beekeeper takes after queen cells containing larvae are found. There is a profound difference between these two conditions. It is not too difficult to prevent a colony from building queen cells; however, once a colony has queen cells with larvae, it is difficult to stop the colony from building more cells and swarming. Therefore, the beekeeper should do all in his power to prevent queen-cell construction; colonies that swarm are useless for honey production in the year they swarm because of their depleted field force.

A queen cell can be difficult for the beginner to define. In its early stages it is called a "queen cup." A queen cup is so called before it has an egg in it or when it contains only an egg. When the egg in the cup hatches, it is then called a "queen cell"; this is also because changes in the cell itself become evident at the same time. About the time the egg hatches the bees begin to add wax to the edges of the cup and lengthen it. Swarm-prevention methods are usually successful when applied to colonies with queen cups containing eggs but rarely successful when applied to colonies with one-day-old larvae in queen cups. Once the larvae are one or more days old, only swarm-control measures are satisfactory.

The presence of queen cells does not always mean a colony will swarm. Swarming and/or queen rearing by a colony can be abortive. However, a colony with queen cells will usually swarm or supersede (i.e., replace the old queen). What starts out to be swarming may become supersedure and vice versa. It is generally accepted that colonies with one to six queen cells are more likely to supersede their queen than to swarm. Colonies with four to twenty cells are likely to swarm. The overlapping figures are intended. The size, shape and position of the cells in a colony also give clues as to what might happen. Generally, when the queen cells are larger, with rougher surfaces, and more on the periphery of the brood nest, swarming is indicated. Smaller cells, built closer to the center of the brood nest, indicates supersedure. These are only general rules. In addition to normal colony variations, certain races of bees build more queen cells than others under apparently the same circumstances.

The swarming season varies as one goes from south to north. Close to the Equator colonies can swarm from February through June, depending upon the food available to the colony. In New York State swarming can occur in May or June, rarely later. While colonies have been known to swarm in August and September, such swarms are infrequent. Usually the presence of queen cells after the normal swarming season indicates the colony is about to supersede its queen.

Clipping a queen's wings is neither a swarm-control nor a swarm-prevention method, nor is it recommended except where a beekeeper is interested in comb-honey production. Clipping a queen's wings may delay swarming but it will not stop it. Usually a colony that swarms leaves the hive within a few days after the queen cells are capped. The bees will attempt to leave even though the queen cannot fly; however, as soon as the bees find that the queen cannot depart with them, they return to the original hive. Often a colony will attempt to swarm several times under these conditions; finally the swarm will depart with the first virgin queen to emerge.

Swarm Prevention

The successful honey producer must build his colonies to a maximum population prior to the honey flow. To do so and to prevent swarming require skill and knowledge in both the art and the science of beekeeping. Proportionally, large populations of bees gather more honey than do small populations; for example, a colony with a population of sixty thousand bees will gather more than twice as much honey as two colonies with thirty thousand bees each. (At the same time, there is no proof that a colony with a population of 150,000 to 200,000 bees will produce more than twice as much honey as colonies half that size. In this regard, a colony with fifty thousand to eighty thousand bees appears to be optimum. This may be true because a queen, at least apparently, is capable of producing secretions in amounts sufficient for about this number of bees.) Larger populations are presumably more efficient in controlling hive temperature, humidity, guarding and other hive tasks than are smaller populations of honeybees. For this reason alone, a beekeeper should take all the steps possible to prevent swarming, for swarming divides a strong colony into two or more units.

While we still have much to learn about the causes and biology of swarming, there is agreement among researchers and practical beekeepers on the primary cause of swarming. The consensus is that swarming is caused by congestion, especially a congestion of bees in the brood nest. The practical swarm-prevention measures used today are aimed at alleviating congestion within this area of the hive.

The best way to prevent swarming is to take the steps that relieve congestion, especially congestion of the brood nest. There are three basic techniques of relieving congestion: reversing; adding supers sometimes accompanied with reversing and/or raising a frame of brood above the existing brood nest; and the Demaree method of swarm prevention (also sometimes used for swarm control).

Reversing, when one has a colony in two supers, is the simplest of the swarm-prevention methods. The first reversing (some beekeepers may reverse three to five times at about two-week intervals before the primary honey flow) usually takes place in early May in New York State. A queen honeybee has a tendency to work upward in the colony as she is laying eggs. Thus, the upper super usually becomes crowded with brood (and sometimes food, too) while the lower super remains relatively empty. By reversing the two supers composing the brood nest, the beekeeper makes space available for more egg laying above the existing (or filled) super. Consequently the queen has additional room to move in an upward direction. Reversing—temporarily at least—splits the existing brood nest.

These two steps—making room available above and splitting the existing brood nest—relieves the immediate congestion and deters the construction of queen cups and, later, queen cells.

The second method of swarm prevention—adding supers—has at least two variations. Often just adding a super above the two existing supers, in early spring (as early as February in the deep south and as late as early May in the north) is sufficient to relieve congestion. Depending upon the position of the brood nest in the super below, this gives the bees the room they need and provides the queen with space for increased egg laying. However, just adding supers may not be enough. Since bees store food above their brood nest, not below it, there is often a barrier of pollen and honey above an existing nest when the third super is added. Therefore, it is usually advisable to reverse the bottom supers at the same time, and to raise one frame of brood into the center of the third super and to place the empty comb it replaces into the center of the brood nest below. While this spreads the brood nest—an act to be avoided if the weather is cool and there is danger of chilling the brood —it is an excellent method of relieving congestion.

The third method of swarm prevention is the so-called Demaree method; it is also a popular swarm-control measure. It has been written about in several journals and texts. Basically the method involves confining the queen in the lowest super with a queen excluder and placing the brood in the third or fourth super above the bottom board. This drastic separation of the queen and brood is perhaps the best of the swarm-prevention methods were it not for the time involved in finding the queen and making the switch of the brood to the upper chamber. Even the best beekeeper cannot make such a manipulation in less than ten minutes and it is not a practical method on a commercial basis; however, for the hobbyist it may be a practical and sound method of colony management. The single danger is that, as a result of this separation, the bees build queen cells in the brood placed in the top super; the beekeeper must inspect the elevated brood to remove any existing cells five to seven days after the original manipulation.

Swarm Control

Once queen cells with larvae are found within a colony in the swarming season, strong measures must be taken to prevent the colony from swarming. Cutting out queen cells is thought of by some as a swarm-control measure, but this is not so. Queen cells are striking and easy to see; however, their presence is evidence, and final evidence, of a more complicated biological process. Cutting out queen cells in a colony can slow swarming, but as soon as the cells are cut and removed, the bees will usually build more. Infrequently a colony will swarm after the cells are removed; this occurs when the cells are cut within a day or a few hours of the time swarming would have taken place normally. The greatest danger in cutting queen cells is that the beekeeper can miss one and so swarming will take place anyway. Comb-honey producers crowd their colonies so as to force the bees to work in the small, crowded confines of a comb-honey section. This encourages swarming. Comb-honey producers cut queen cells, when they are present, every seven or eight days. They also clip the wings of the queens in their colonies so that if a swarm emerges, it will return to the parent hive. If a comb-honey producer misses a queen cell under these conditions, the swarm will leave with the virgin queen when she emerges. Even comb-honey producers find it is not profitable to remove queen cells more than about three times; after the third time—and colonies are carefully marked in this regard—a colony that persists in constructing queen cells is requeened or used for other purposes.

The presence of queen cells is only one manifestation of the swarming process. Research has shown that many other things occur in the colony at the same time. For example, queen bees lose about one-third of their weight during the four to five days prior to the time the swarm emerges from the hive; this weight loss is necessary so that the queen may fly, for otherwise she is too heavy to do so. Such a weight loss obviously affects egg laying and other queen activities. There is increasing evidence to show that foraging slows during the several days prior to swarming; large numbers of bees engorge in preparation for swarming and this, too, occurs several days prior to the time the swarm actually emerges. Thus, swarming does more than divide a colony; it robs the colony effort at a time when having a strong, populous colony is necessary to gather honey.

When queen cells are found in a colony, there are only three basic methods of swarm control, each with some variations, which may be used to prevent swarming: removal of the queen, removal of the brood, and separation of the queen and brood. One can also split the colony into two parts, requeening the queenless half or allowing it to rear its own queen. However, this is nothing more than artificial swarming and creates two weaker units that will not be able to produce as much honey had the colony not been divided.

Separation of the brood and queen is the Demaree method, described above, and can be used in swarm prevention or swarm control. One popular variation of this swarm-control technique is to give the brood above the old colony a new queen, thereby creating a two-queen colony. The two units are later combined, just before the honey flow, so as to produce a populous colony for honey production. The old queen can be killed before the two units are united, or the two can be put together, allowing the two queens to fight; in this case the younger queen usually survives. There are also several variations of this so-called two-queen system of management. Generally speaking, this method is too time-consuming for use in the northern United States; also, there is often a serious food shortage between late May and mid-June, after the dandelion and yellow rocket honey flows and before the major honey flow begins in June. In certain years heavy feeding of colonies is required, especially of those colonies which have large populations of adults and brood. Two-queen systems have been described in several journals and bulletins and for the hobbiest may be worthwhile and interesting.

Removal of the brood, or of the queen, is likewise a drastic step, which if not properly done can weaken the colony so much that a surplus of honey will not be gathered. Dr. C. C. Miller, whose books on comb-honey production contain the best bee-management information available, recommends caging a queen for a week or ten days to control swarming. This has the same effect as removing the queen. This interruption stops egg laying and relieves congestion. At the same time it usually prevents the construction of additional queen cells, though it is still necessary to cut out those already built.

Depending upon the strength of the colony, it may be necessary to remove only a few frames of brood and to replace these with empty combs in the center of the brood nest. This step, too, must be accompanied with cutting out the cells already built. It involves a separation of the brood nest—a dangerous practice if cool weather prevails, for brood can be chilled and killed. The person who successfully practices the art of beekeeping takes the necessary steps at precisely the right time. Experience is the best guide. No two colonies, no two seasons are exactly alike; this provides a constant challenge to the beekeeper in his efforts to maintain strong colonies for honey production.

Supering

Giving colonies extra supers of combs before the honey flow aids in swarm control; there are other advantages to the beekeeper, too. While most beekeepers can judge when a honey flow will take place, it is not always possible to do so. For this reason alone the beekeeper should always make certain his colonies have adequate room for nectar storage. Additionally, supers of combs that are on colonies will not be attacked by wax moths or other pests, provided the colonies are sufficiently strong.

When the beekeeper wants liquid (extracted) honey production, he should add new supers on top of the hive. This means adding empty supers on top of already filled supers and is called "top-supering." When comb honey is being produced, the new comb-honey supers are placed under the already filled or partially filled supers and immediately on top of the brood nest. This is called "bottom-supering" and is a basic difference in colony management between the two types of honey production. It has been argued that bottom-supering is helpful in liquid-honey production because it places empty comb close to the brood nest and in a position where it is readily accessible. This is probably true, but most beekeepers do not consider it too important and the lifting involved is hard work.

Routine Maintenance of the Apiary Site

If a beekeeper has only one apiary, and it is adjacent to his residence, the site is usually mowed and tended with the rest of the home grounds. There is one note of caution about mowing around bee hives with mowers with gasoline engines. For some reason the noise or more probably the odor from a gasoline engine irritates the bees and they are prone to fly out and sting anyone pushing the mower in front of a hive. It is a good idea not to work the colonies while or immediately after there is mowing at the apiary site. It is advisable to mow the grass around colonies early in the morning or late in the evening, when the bees are not so active and not so easily disturbed.

Keeping the grass neat around a colony can be facilitated by placing a piece of protruding tarpaper or roofing paper underneath the hive stand. If the tarpaper protrudes two to four inches, one can mow around the hive stand with ease and there will be no accumulation of grass there. Likewise, the tarpaper will prevent the grass from growing up between the colonies on a hive stand.

Beekeepers with remote apiaries find that they usually mow them less frequently. The grass in front of the colonies should be prevented from growing. This can be done by using leftover black winter wrapping paper, which can be tucked under the hive stand and can extend out from the colony two or three feet and be held in place with stones. Other beekeepers have used salt to prevent the growth of grass around colonies in apiaries; however, large quantities of salt are required and it is probably not a practical way to deal with the problem. Still others have found that weed killers can be used without any adverse effects on the bees, especially if the weed killers are used early in the spring, at which time a low dosage is needed to kill the grass around the hive. As is the case with insecticides, small quantities of pesticides used around hives when the bees within are not active, will not have an adverse effect on the bees provided the pesticides do not actually come in contact with the entrance of the colony or contaminate the interior of the hive. If the growth of grass around colonies is not somehow controlled, the colonies will become shaded and flight to and from the colonies will be impeded.

IV

Summer Management

It is during the summer that the beekeeper reaps the rewards of his efforts for good fall, winter and spring management. Summer management of honeybees is often more hard work than application of good judgment; in the summer the colonies are heavy with honey (or at least they should be) and many of the practices are routine and involved with securing the crop. Still, there are certain items that should receive attention as apiaries are visited and colonies inspected. These are discussed in this chapter.

The Primary Honey Flow

In most areas in the northern United States the honey flow from clovers and alfalfa, starting sometime in mid- or late June and lasting usually through July, is the flow on which the most and the best-quality honey is made. Thus, it is called the "primary honey flow"; for a profitable commercial beekeeping operation the primary flow should produce a surplus of 100 pounds per hive. In recent years, in certain parts of the northeast, the honey flow from alfalfa has been most important and the primary honey flow. Each honey-producing area in the country has at least one primary honey flow, though in some areas there may be a second honey flow as good as the primary one. The term "primary honey flow" is especially helpful when one thinks about management schemes.

Keeping records of honey flows over a period of years will help to establish which honey flow is most important in a given area. At the same time beekeepers must be aware of changing conditions in their area. In some parts of the north, goldenrod, which starts to yield nectar in August, is of prime consequence. This is especially true where clover is no longer grown in quantity for dairying. In such areas it is often advisable to split the colonies, early in the year, into two or three units, planning to build these to maximum strength by about the first of August. Thus, there is no exact management scheme for the primary honey flow. It varies from area to area and, to some extent, from year to year. Record keeping is the best way to estimate when and what should be done in future years; a hive on scales is probably a good investment for most beekeepers.

Queen Excluders

Some beekeepers feel a queen excluder is indispensable in their management system; others feel it is a nuisance and/or a honey excluder, which may interfere with normal ventilation in the colony. It is the author's opinion that a queen excluder has more advantages than disadvantages, if properly used. Excluders work well in cases where bee escapes are used too, as well as in other management systems.

Queen excluders should not be put on colonies any earlier in the season than necessary; they seldom serve any useful purpose before July 1 in the north. Excluders should be removed as soon as possible in the fall; this is usually done when the fall honey is removed. While some beekeepers put the queen excluder above the second super, it is best to put the queen into the bottom chamber and to place the excluder above this single chamber. There are several reasons for doing so. Usually by July 1 in the northern states the bottom brood chamber is empty. Thus, if the queen is forced to lay in this super she normally has plenty of room. (A standard frame contains 6,000 to 7,000 cells; since a queen can lay a maximum of about 1,500 eggs a day obviously she can be forced to do all of her egg laying in one standard super.) Secondly, if the queen is prevented from laying in the second brood chamber, the bees will fill it with honey, and when the colony is wintered, it will have a full super of good-quality honey for winter food. Usually the quality of the honey gathered in July in the north is superior to that gathered in August or September; it is therefore better for winter food.

The chief question about excluders is what is the easiest and most efficient way to put the queens down; certainly it is too time-consuming to search for and to find the queen and to place her below the excluder. This method may be practical for one who has only a few colonies but not for the beekeeper who has more.

There are basically three methods of forcing the queens into the lower chamber: (1) driving the bees down with a repellent; (2) smoking the colony heavily and forcing the bulk of the bees and the queen down; and (3) shaking the bees and the queen off the brood combs, either in front of the colony or into the bottom super. Sometimes a combination of these methods is used. None of these methods is perfect and if many colonies are examined, one is almost certain to leave the occasional queen above the queen excluder. Of all these methods, probably using a repellent to drive the bees and queen down is easiest and most efficient. The repellent board is used just as it is when honey supers are removed, except that one starts at the top of the brood nest, not at the top of the hive, since the queen remains in the brood nest. A danger in shaking is that the queen may be injured.

How Many Supers to Add

At the end of the honey-producing season, when the crop is ready to be removed, the best colonies will be in four to seven supers. This does not mean that they will have two to five supers above the double brood nest full of honey; it merely means that they have some bees and some honey in this number of supers. When a beekeeper produces liquid honey, he expects that not every comb will be completely filled with honey. This is quite different from comb honey production; when making comb honey the colonies must be extremely crowded.

When one produces liquid honey, it is best to have more than enough supers on the colonies. There is no absolute rule to follow and this fact is one of the great fascinations in beekeeping. Generally it will be observed that beekeepers will have one or two more supers on their colonies than the bees actually need in the average year. This means, too, that occasionally one can expect to fill all the supers and to produce a bumper crop.

Adding Foundation and Making New Combs

It is good beekeeping practice to make a few new combs each year. Not only do cells in the comb become smaller with each cycle of brood because of the accumulated pupal skins (cocoons), but also combs become worn and broken with age; old combs often contain too many drone cells. Combs not properly wired can sag and stretch the cells.

The best time to draw foundation or to make new combs is during a honey flow. While it is true that bees can draw foundation almost any time during the active season, especially if they are fed sugar syrup, they do the best job during a flow. The best place in the hive to put new frames with foundation is just above the brood nest. Probably the ideal situation is to place six new frames with foundation in the center of the second super with an excluder on top and the queen below. The outside combs, those in addition to the six new frames, would be drawn comb.

The second best way to draw foundation is to place four new combs, spaced alternately between five drawn combs, or to place an entire super of foundation immediately above the super in which the bees are storing nectar at the time the new combs are being put on the hive. Some beekeepers make a practice of adding one or two new combs to every, or almost every, super as they are given during the honey flow.

The reason that new foundation should be added during the honey flow is that bees have a tendency to chew the edges of the wax foundation if nectar is not readily available. This is one of the reasons, too, that sections for comb-honey production are not added until the honey flow is underway.

Great care should be taken in extracting new combs. It is best to remove much of the new cells when uncapping new combs. Often the uncapping knife is forced to cut so deep that only one-quarter to one-half inch of cell wall, in depth, is left. In this way much of the weight is removed and there is less danger of breaking the comb in the extracting process. When new combs are extracted, the machine should be run very slowly so as not to put unnecessary pressure on the comb. Comb protectors can be purchased for use in extracting new frames, but they are probably not necessary if sufficient care is exercised.

V

Removing the Crop

As soon as the honey flow is finished and the combs are filled with new honey, they should be removed and the honey extracted. One reason extracting should not be delayed is that certain honeys crystallize in the comb and once they do so, it is extremely difficult to remove the honey. Some of the honeys produced by fall flowers, especially aster, are notorious for their rapid crystallization rates. It is not infrequent that aster honey will crystallize within ten to twenty days after being stored in combs.

In the northern United States most beekeepers extract their honey twice a year, usually first at the end of the light or clover honey flow sometime in late July or early August, and then again following the fall flower flow sometime in late September or October. Honey from fall flowers is usually darker and stronger in flavor. If the beekeeper is interested in different types of honey, or in keeping the higher-quality clover or early honey separate from the rest, an early extracting is required.

In the case of comb honey the sections may become travel-stained by bits of pollen and propolis from their feet as bees walk over the comb surface. Comb-honey sections lose much of their attractiveness if they are not removed as soon as possible. Then, too, bees may store nectars of varying colors in different cells in the same comb and often this gives the section a poor appearance.

Methods of Removing Honey from Colonies

There are several ways of removing supers of honey from a colony of bees. These include brushing or shaking the bees from the combs, placing bee escapes under the full supers, driving the bees with repellents, and the use of forced air. Each of these methods has its advantages and disadvantages and there is no agreement in the beekeeping industry as to which is actually best.

Since honey is usually removed after the honey flow, and at a time when only a limited amount of nectar may be available, robbing may be stimulated. During a dearth of nectar just the odor of exposed honey is apparently enough to trigger robbing. At the time the combs and/or supers are being removed from the hives great care should be taken to keep all the new honey covered or screened so that robbing will not be encouraged.

The Bee Escape

The Porter bee escape was invented in 1891. The bee escapes now on the market fit into the hole in a standard inner cover. Inner covers with bee escapes in place are technically called "bee escape boards," though beekeepers speak of them as "bee escapes," or sometimes just "escapes." Although the bee escape is an excellent way to remove honey from colonies of bees, there are two disadvantages. The first is that two trips must be made to the apiary, one to put the bee escape in place and the second to remove the honey; and secondly, if there are holes in the supers above the bee escape, the bees may rob the honey from the supers. However, despite these two problems, the use of bee escapes remains a popular method of removing honey supers and is extensively used by both hobbyists and commercial beekeepers.

A knowledge of bee behavior is helpful when bee escapes are used. If there is dripping honey above the bee escape, the bees will be more reluctant to leave the supers. Therefore, if one is removing more than one super of honey above an escape, it is helpful if the individual supers are not broken apart. Usually two men are required to lift the supers of honey with hive lifters, while a third man inserts the bee escape. However, experienced beekeepers have also found they can break the hive apart at the point where they wish to insert the bee escape by using a hive tool, partially put the escape in position and, holding the supers with their hands, push the escape into place with their stomachs.

Experienced beekeepers with equipment that contains holes and/or cracks through which bees can rob honey have found that, if they put the escapes on early, or in midmorning, they can remove the supers the following morning before any robbing can start. However, this is possible only if there is no dripping honey or broken combs above the escape that might delay the departure of the bees. If one delays in removing supers to which other bees can gain access, the robber bees can remove all of the honey within a few hours. At the same time, one cannot expect bees to abandon brood. If there is brood above a bee escape, some worker bees will remain with it; for this reason many beekeepers who use bee escapes also use queen excluders.

Bee escapes can be left in place for two days or more, usually without difficulty. Robbing can be a problem. There is also a slight danger of the combs overheating and melting during very warm weather above a bee escape because there is no ventilation. All the advantages and disadvantages must be weighed by the prospective user. One great advantage of the use of a bee escape is that beekeepers with many colonies may start very early in the morning, before the bees fly, to remove their honey, and do so without veils or gloves. Once the bees have been removed the honey supers can be removed from the colonies with ease.

Bee Repellents

Certain chemicals repel honeybees and drive them off combs of honey. The first material used commercially in this regard was carbolic acid, which was used from the early 1930's to the early 1960's. However, carbolic acid has certain undesirable characteristics and its use is not approved by health agencies. Two materials, propionic anhydride and benzaldehyde, have replaced carbolic acid, with the latter being more popular. However, neither material is perfect; the one who discovers a perfect repellent, which would work well under all conditions, would help the industry considerably.

Repellents are used by placing them on an absorbent pad made of cloth or other material that is placed on a wooden rim the size of a honey super but only two or three inches deep. It is advisable to start the downward movement of the bees with smoke before putting the repellent board in place; otherwise, bees on or near the tops of the combs can become confused by the repellent and move in the wrong direction. Repellents are affected by temperature and work more rapidly at high temperatures. On very warm days the quantity of material needed to drive bees off combs in supers will be much less than is needed on a cool day.

Brushing and Shaking Bees

The oldest method of removing combs of honey from a colony is to remove them one at a time, after having smoked the super, and to shake and brush the bees off the combs. This technique works well enough if one has only a few supers of combs to remove. However, bees that are brushed and shaken can become quite angry. The chief advantage of the method is that no special equipment is needed. A brush of grass works quite well. Shaking and brushing bees involves more exposure of the combs of honey than any other technique. It is therefore quite easy to stimulate robbing at the time and special care should be exercised to avoid this.

Forced Air

The use of forced air to blow bees out of honey supers has become increasingly popular in the past few years, especially with some of the larger commercial beekeepers. Since elaborate equipment and some type of power supply is needed, the method is not too practical for the beekeeper with only a few colonies. Still, some beekeepers have built homemade blowers that work quite well. The physical act of blowing the bees out of the supers apparently does them no harm.

Extracting

Extracting is the process of removing liquid honey from the comb. The main reason that honey is extracted is so that the comb, which is costly both for man and the bees to produce, can be reused. Insofar as the beekeeper is concerned, the object should be to remove the honey and retain as much of its quality as possible. All too frequently the honey is improperly extracted or damaged in the extracting process. The major problems are incorporating too much air, which results in foam on the top of the final package, and overheating the honey before it is bottled.

The extracting process involves three basic steps: uncapping the combs of honey; placing the uncapped combs in the centrifugal-force machine (extractor) to remove the honey; and giving the honey a straining to remove any bits of wax or other

A frame of honey, almost completely capped, removed from a colony in Florida. This standard frame contains six to seven pounds of honey; the honey is removed from frames by centrifugal force (with an extractor). The comb is reused, thus saving the bees the effort of building new comb.

extraneous material that might be accidentally introduced into it. The extracting process can be speeded up considerably if the honey is warmed before it is removed from the comb. Some beekeepers who have only a few colonies of bees extract the honey immediately after removing it from the hive, thus taking advantage of the high hive temperature to keep the honey warm. Commercial beekeepers often place supers of honey in a heated room, a room usually kept at about 95 degrees Fahrenheit. It is also possible to force air through supers, both to remove moisture and to warm the honey.

Several types of uncapping knives and machines are available. A beginning beekeeper, with only a hive or two, will be satisfied to use a so-called cold knife. This is a knife not too unlike a butcher knife, with a fairly heavy steel blade. The knife is placed in hot water so the blade is warmed prior to cutting the wax cappings from the comb. Usually two knives are used; one is heating while the second one is used to uncap. Beekeepers with more colonies may care to use electrically-heated or steam-heated knives. Some commercial beekeepers use power uncappers, which remove the capping from many combs per minute.

As is the case with uncapping methods, there is a great variation in the size and type of extractors available. The beginner will find that a two- or three-frame nonreversible extractor will suffice; extractors are available that hold as many as 50 combs.

Since equipment for extracting can be expensive, it is often advisable for beginning beekeepers to cooperate with one another or to seek out a commercial beekeeper who will extract the honey at a rate of so many cents a pound. This has the added advantage of acquainting the beginner with the various types of uncapping knives and extractors on the market together with their advantages and disadvantages. Often second-hand extractors and uncapping knives can be purchased at the same time one is purchasing secondhand beehives. Doing so might save considerable investment.

Cappings

The wax cappings that are cut from a comb to be extracted usually contain quite a bit of honey. This honey can be salvaged, but not without some difficulty. Cappings usually retain honey that can be removed only after they are crushed and broken.

Perhaps the easiest way to remove honey from cappings is to place them in a wooden or metal container with a coarse screen bottom and to allow them to drain for a day or two. Some beekeepers place such containers in a special warming oven or room since warming the cappings will speed up the draining process. If the cappings are broken with a knife or trowel, they will drain reasonably well.

Cappings can also be placed in an extractor and the honey centrifuged out in much the same way it is removed from a comb. However, this is a sticky process and allowing the cappings to drain is easier even though it does take a longer period of time. Commercial beekeepers have a variety of ways to remove the honey from cappings. One old method is to place the partially drained cappings in a wax press and press the honey out. More recently a continuous centrifuge has been developed that can remove the honey from cappings; however, this is a costly machine used only by beekeepers with many colonies. A few beekeepers who have apiaries in remote locations, where there are no other bees, sometimes use such a site to allow the bee to rob the cappings, thus further salvaging honey and at the same time making it easier to handle the cappings. Cappings that have been robbed and are not sticky with honey may be shoveled or handled much as one would treat grain.

Since cappings are at the most only moderately travel-stained and contain little or no pollen or propolis, capping wax is lighter in color than is wax rendered from old combs; for this reason capping wax usually receives a premium in the market. It is advisable to keep capping wax separate from old comb wax when the wax is rendered or sold.

Wholesaling Honey

The beekeeper who produces more honey than he can use himself or sell in the immediate vicinity may care to package his honey for the wholesale trade. The two common bulk-honey containers are the 60-pound can and the 50-gallon drum; however, the 60-pound can is not so popular with large honey packers today. Some honey packers pay less for honey in 60-pound cans because more labor is involved in handling this smaller container.

Beekeepers who produce only a few drums a year find they do not necessarily need special equipment to handle these large containers. If the honey in the drums is allowed to crystallize, the drums can be tipped on their sides on an old tire. Following crystallization, they can be rolled without danger of spilling the honey. Most honeys crystallize rapidly in the fall. Crystallization can be encouraged by stirring into the drum some already partially crystallized honey.

Honey packaged in large cans or drums can ferment if it is not pasteurized or if the moisture level is too high. Few beekeepers pasteurize the honey they sell to the wholesale trade. They do attempt to extract only honey with the proper moisture content and/or to force crystallization, which may partially eliminate fermentation. Storing the honey in an unheated warehouse in the winter, at a temperature well below 50 degrees Fahrenheit, will protect the honey from fermentation. Honey sold in the wholesale trade should be strained so as to remove all the excess wax particles. However, there is seldom any price penalty for honey with wax particles if the grades are U.S. B or C.

Small-Scale Processing of Liquid Honey for Market

The beekeeper who produces only a small quantity of honey is at a disadvantage in packing his product. The large honey packer is able to blend to suit the taste of a market. He may also blend his honey so as to balance the moisture content, or, if necessary, use a vacuum pan to reduce the moisture content of high-moisture honey. Most large packers of honey filter it, which gives it a distinct sparkle and also extends the shelf life. While filtration has these advantages, we know, too, that it removes the pollen and a certain amount of the flavor from the honey; the small beekeeper might be able to use this fact to advantage in his advertising.

Small producers should do the best job possible to sell a product that will result in repeat sales for himself and other beekeepers. In many areas of the country it may be advantageous to buy a small quantity of honey and blend it with that which is being packed for the local market. When a honey with a distinct flavor is packed, a special descriptive label, or a second label, can be used to advantage.

Since the beekeeper has little control over the moisture content of his honey, it is best to remove from colonies only honey that has been fully ripened by the bees. It is possible to remove moisture from honey still in the comb by placing it in a heated room and directing a current of warm air over the combs. In fact, in a well-made ventilated room, it may be possible to remove as much as 1 percent moisture from honey in the comb in twenty-four hours. However, it is almost impossible for the small beekeeper to remove moisture from honey once it has been extracted from the comb.

Bottling and Labeling

The standard retail packages in which honey is sold are the one-half-, one-, and two-pound jars and the five-pound tin or jar. Liquid honey prepared for the retail trade should be strained or be as clean as honey strained through an eighty-mesh screen. Honey that has been heated and allowed to settle or that has been baffled will usually be as clean as that which is strained. The honey should be heated to 140 degrees Fahrenheit for 30 minutes or 160 degrees Fahrenheit for one minute or some intermediate combination. Honey packed in the jar and capped while it is still hot will have a longer shelf life (will not granulate so rapidly) than that which is packed cold.

Most beekeepers who pack their own honey use a jacketed hot-water tank heated with electricity or gas. Several models are available from the bee supply companies. Too often the honey is overheated, usually because it is kept at a high temperature for too long before being packed.

In commercial packing plants a machine for removing dust from the new glass jars is used before they are filled with honey. Even jars just received from the factory may contain dust, which can cause premature granulation. For this reason the beekeeper should make some provision to clean the jars before the honey is packed in them.

Labels for honey jars are available from the bee supply companies. All too often the label is designed to sell to the beekeeper, not for the beekeeper to sell honey to the consumer. Most beekeepers like to see a picture of a bee or a beehive on their honey label; most consumers think of bees as just another insect and the picture of a bee is not so appealing to them. Certain colors complement honey more than others; which colors should be used on the label depends upon the color of the honey. Most of these questions concerning market-ing and the factors that affect the sale of food products have been carefully researched. The marketing-minded beekeeper may care to pursue the subject in greater detail.

Roadside Stands

Some beekeepers have used roadside stands effectively to sell all or a portion of their honey crop. The chief advantage is that the beekeeper sells his honey at the higher, retail-market price. The major disadvantage appears to be that customers may talk so long or ask so many questions about bees that the beekeeper loses too much time when making a sale.

Self-service honey stands are popular with many beekeepers; they are equipped with a box, open or with a sealed cover, into which the customer deposits his money. One advantage of a self-service stand is that the beekeeper's time is not wasted. In areas of high population density thievery is some-times a problem. At the same time a roadstand is effective only in a well-populated area or where there is heavy traffic on the road.

Roadside stands, servicing of garden stores and vegetable stands, the mail-order trade and wholesaling are all possible methods by which a beekeeper can sell his honey. The author has known beekeepers who have done well in each of these areas; a method of selling honey that is used successfully by one person may not be so successful in the hands of another. The beekeeper, like all persons in agriculture, is at a disad-vantage as he must sell his product at the market price. While producers of farm products claim to be independent men, in actual fact they are at the mercy of a worldwide market, much of which is controlled by federal government pricing both in this country and elsewhere.

Variation in Honey

Just as flowers vary in size, color, shape and odor, so the nectars they produce vary. Bees use all these factors for orientation as they fly from one flower to another in the field. Man benefits from this great variation and harvests honey with many colors and flavors. This has both advantages and disadvantages for the beekeeper. He, and those who understand these variations, will enjoy tasting different honeys; the customer who buys only one or two jars of honey a year can be confused when the second jar of honey does not taste like the first.

One cannot expect the customer to know everything about the product he is buying. For this reason it is often advisable to label the jar with the type of honey it contains. Some beekeepers have a second label printed with descriptive information; this label is placed on the jar opposite the first one.

Then, too, there are probably certain honeys that should not be placed on the general market simply because they are too strong and not known by the average consumer. Often certain of the stronger-flavored honeys are better appreciated if they are blended with a milder honey. The flavor of the strong honey still dominates but is not so harsh as to be objectionable. Often, in his enthusiasm to produce and market a certain type of honey, the beekeeper forgets that the public has little or no knowledge of honey. In general the American public prefers a light-colored, mild-flavored honey. The successful seller of honey must cater to these tastes; he may produce and market other types of honey, but this, too, must be done with care.

VI

The Fall Crop
and Fall Management

Sometime between August 1 and mid-October, in much of the north, goldenrod may produce a surplus of nectar. This may be followed by a honey flow from aster, but aster does not yield nectar in quantity until after a frost hard enough to kill the goldenrod.

Soon after the first of August is the time the beekeeper should begin to think about the general condition of his colonies. A good colony of bees should have a young, vigorous queen that will lay eggs late in the fall so as to have as many young bees in the winter cluster as possible. A young queen is also an advantage in the spring as such a queen will start brood rearing earlier in the year and produce a more populous colony by the time of the June-July honey flow. For these two reasons it is often said the beekeeper's year begins August 1 and that, where annual requeening is practiced, it should be done at about that time.

Swarming and Fall Management

Bees seldom swarm in the fall. The beekeeper need not worry so much about providing ample room for the brood nest after about the first of July; however, ample room should be made available for honey storage.

The author prefers to place queen excluders on colonies above a single brood nest super sometime after July 1. Many beekeepers argue against using queen excluders, but excluders make fall management easier. Since crowding the queen into a single super after July 1 seldom precipitates swarming, there is little danger in this regard. In fact, during the first two weeks after the queen is placed in the bottom super, congestion is usually relieved; usually the bottom super is empty and contains little or no brood or honey or pollen at this time of year. It has been asserted that queen excluders hinder the movement of bees in the colony and interfere with ventilation of the hive, but there are no statistical data to back up these statements or to prove that queen excluders have an adverse effect on honey production.

The chief advantage of using a queen excluder above a single brood nest super is that the bees will fill the second super—the one immediately above the brood nest—with honey. In the north a colony of bees needs sixty to eighty pounds of honey to winter successfully. The preferred system of wintering bees involves using two supers and it is important that one of these is packed full of honey. If a queen excluder is not used, the queen will often continue to use at least a part of the second super for brood rearing. This means the beekeeper must rearrange the combs in the supers in the fall, a job that is not impossible, but easier left to the bees.

A second advantage of a queen excluder is that one need not worry about having brood in combs about to be extracted.

Colonies can be moved from one location to another without difficulty; this top screen, which provides ventilation, is held in place with four nails.

The entrance of a colony to be moved can be closed by tucking in a piece of ordinary metal fly screening about two inches wide.

Loading single-story colonies on a truck for transport from New York to Florida in the fall; the colonies will be returned in the spring. Thousands of colonies are moved south in the fall and north in the spring from several states across the country.

Where bee escapes are used, excluders are also an advantage.

In most of the northern states fall honey flows are more erratic than spring honey flows. Goldenrod, of which there are many species, does not yield nectar every year or in every location. When goldenrod does yield, crops of fifty to over one hundred pounds per colony are not unusual.

The author once thought he had devised a perfect system for the management of bees in the southern tier of New York State. The aim was to make a maximum fall crop, which is the chief honey flow in this area. The overwintered colonies were split and the queenless halves requeened, or allowed to requeen themselves, in late April. This circumvented the swarming problem since the colonies were too weak to swarm in May and June. The positions of the colonies were rotated in the apiary in June and July to equalize their strength. By August 1 the colonies were usually in three or four supers and had a small reserve of honey. On about August 1 the queens were driven into the bottom super using a repellent. An excluder was added above the first super. The colonies were then given an additional super or two.

The goldenrod flow in the north can start as early as August 1 or as late as about September 10. Occasionally there is no honey flow at all, and this is when the system fails.

In a good year the goldenrod will be frost-killed in late September and this may be followed by a honey flow from aster. When it is time to remove the honey, using the above system of management, it is only necessary to select those colonies which are in the best condition for wintering. The second super, which should weigh 60 to 70 pounds, is left for the winter food chamber and all the supers above this are removed and extracted. The remaining colonies may be killed and the entire crop removed. The honey yield in a good year, based on the spring count, will usually be above 200 pounds, but only in those years when there is a good goldenrod honey flow.

Requeening

A young queen lays more eggs, maintains a more compact brood pattern, usually has a stronger colony that is more resistant to stress diseases, lays eggs later in the fall and earlier in the spring, and finally and perhaps most important, produces more of the secretions important in maintaining social order in the colony. For this last reason, swarming is much less of a problem when a young queen is present in a colony. Annual requeening of colonies is a practice recommended in most texts on bees. It is routinely practiced by some commercial beekeepers; other beekeepers try to watch their colonies more closely and attempt to requeen only those colonies which have failing queens.

When annual requeening is done, it is probably best to do so about August 1. September requeening may be satisfactory, but not all colonies will accept a new queen. If the requeening is done in early August, then it may be possible to requeen those colonies in which it first fails.

There are several methods of requeening, some easier and more certain than others. A common method is to find the old queen and kill her, placing a queen cage with the young queen in the brood nest to be released by the bees in a day or two. How well this method works depends upon a number of factors. If there is a honey flow in progress, the queen is more readily accepted, though even this is not always true. This method of requeening is not recommended for beginners and those not familiar with bee behavior.

A preferred method of requeening is to introduce a young queen into a nucleus colony containing one frame of brood and two or three frames of bees about mid-July. A young queen is almost always accepted under such circumstances. About August 1 the queen in the old colony may be found and killed.

A single sheet of newspaper is placed above the old queen's brood nest and the nucleus colony placed above the newspaper. The honey supers are placed above this, but if there are too many bees in them, they should be shaken out, or a second sheet of newspaper placed between the nucleus and the honey supers. By the time the newspaper is chewed away, the colony odors will be the same and the young queen will be accepted. If a queen excluder is to be placed on the colony, this should be delayed for a week or two, at which time the young queen may be successfully driven into the lower brood chamber.

It is good beekeeping practice to keep a few nucleus colonies in reserve for emergency requeening. Whenever two colonies are combined the safest and easiest method is to use the single sheet of newspaper; most beekeepers cut a few small slash lines in the paper to speed up its removal by the bees, but this is not really necessary.

The Quality of Fall Honey as Winter Food

In the winter honeybees cluster. Not only do they fill the spaces within the comb, but many crawl into cells. The outer shell of this cluster resembles a ball in shape, and like a ball, the outer surface is a compact mass (of bees) while the inside is hollow. A cross section of what a winter cluster might look like within a colony may be seen in an observation hive on a cold fall or spring day if the hive is in a cool room.

Bees in the center of the cluster warm the cluster by eating honey and physically moving to generate heat. Large quantities of honey are necessary to keep the colony alive throughout the winter. Honeybees do not defecate within the hive under ordinary circumstances; in fact, if they do so, social order usually breaks down completely and the colony dies.

Since bees consume large quantities of honey in the winter and waste products accumulate in the lower digestive tract as a result, it is important that either the food be as free of in-digestible matter as possible, or that the bees have frequent flights to void fecal matter. Generally speaking, fall honey contains more undigestible matter than the lighter-colored honeys producer earlier in the year.

Today only a few beekeepers winter their bees in cellars, but those who do follow the practice of feeding their colonies sugar syrup in the late fall, a practice long advocated by those who have used cellars extensively in the past. Beekeepers in the more northern parts of the northeast, but who winter their bees outdoors, can also feed the colonies ten or twenty pounds of sugar syrup in the fall, or they can attempt to use only high-quality honey for winter food.

As mentioned earlier, some honeys have a tendency to granulate more rapidly than others. Aster honey makes very poor winter food because it granulates so hard that it is difficult for the bees to remove it from the cells in the comb. Thus, the question of winter food is more important in some areas than in others. It is important for the beekeeper to understand his situation locally and to take the steps necessary for the successful wintering of his colonies.

VII

Wintering Bees

In the northern United States bees should be given some special protection during the winter months. While a colony can survive without special attention—and many do—the general condition of the colony will be improved if they are given some packing. Cold weather is not a problem for honeybees; their efficient clustering system is designed to protect them against great extremes of cold. However, in the process of metabolizing honey, bees give off great quantities of water. It is important to get rid of this water and that it not condense within the hive. At the same time, bees consuming honey accumulate fecal matter, which they must void outside of the hive if the colony is to remain in good condition. In the case of older bees, it is far better that they are encouraged to fly out of the hive in the winter to void feces, and not return, than it is that they remain in the hive. The use of winter packing is designed to assist the bees in these two problems of ridding the colony of excess moisture and of encouraging winter flight as much as possible so that the bees may rid themselves of feces.

Weighing Colonies

Few beekeepers weigh their colonies in the fall, but it is the only accurate way to determine how much food a colony has stored for winter. Most beekeepers merely lift their colonies from one end to check on winter stores; this is not a satisfactory method of checking the quantity of winter food and is discouraged. Buying a pair of scales for weighing colonies is well worth the investment.

A two-story colony with an ordinary bottom board and cover, combs and bees should weigh 130 pounds to winter satisfactorily. If the colony weighs less than this, it should be fed. Feeding should be done shortly after the goldenrod is killed in the fall. By feeding at this time of the year the bees have time to arrange their winter stores in the manner they prefer.

There are a variety of ways of weighing bees, but for one man a two-legged stand is probably most satisfactory. Scales mounted on a pole for weighing by two men works rapidly. The weights may be written on the colony so the beekeeper knows exactly how much syrup to feed.

Fall Feeding

Bees dilute honey with water before they consume it; however, when nectar is stored, moisture is removed from it. For these reasons sugar syrup that is fed for winter storage should be more concentrated than that which is used for spring feeding. In the spring the syrup is made with one part sugar and two parts water; in the fall the mixture is made one to one by weight or volume.

Bees cannot digest all of the sugars man can. This is seldom a problem as most beekeepers feed their bees only sucrose or ordinary table sugar. If some other sugar is used, the literature should be checked to make sure the bees can digest it. Waste sugar, or sugar that is salvaged in candy plants, etc., is sometimes fed to bees. This is better used as spring food as it may

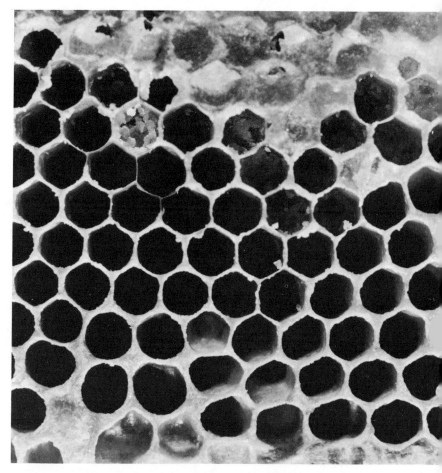

A frame taken from a winter cluster of bees in January shows empty cells below into which some bees crawl to form the outer surface of the winter cluster. The capped honey cells above would soon be opened for winter food. The cells in the center have been opened and the liquid honey has been removed. The bees have left the crystallized honey, which they are capable of dissolving only slowly; because this honey formed hard granules, it is not good winter food for bees.

contain some indigestible matter. Only a high-quality food—that is, one which contains only sucrose—should be given to bees in the fall.

There are many ways to feed sugar to bees: pails, division board feeders, entrance feeders, pouring syrup into empty combs, pan-feeding above or below the brood nest, dry-sugar feeding, and mass feeding outside of the hive. (See "Spring Feeding," page 53.)

Methods of Packing Bees

In the northern states a variety of methods have been used to pack bees for winter. (The term "pack" is a misnomer and dates from the time when bees were "packed" with sawdust or straw in a refrigerator-type packing box. The theory at the time was that bees needed assistance to conserve heat; we now know that this is not true.) Winter preparation may or may not mean placing a protective wrapper around the colony, depending upon the severity of the winters.

The most popular and successful winter wrapping material is black building paper. The lighter the weight of the paper, the better. Many beekeepers use slater's felt for winter wrapping, but this is too heavy. For the most successful winter pack the paper should be waterproof but allow some slight movement of moisture-laden air.

Several years ago wooden packing cases were commonplace. A few beekeepers have made metal packing cases. Research on the winter survival of honeybee colonies indicates heavy winter packing is not helpful. The winter cluster must be able to move to new stores and the bees within the hive must be able to rid themselves of accumulated fecal matter. These two factors are more important than anything else insofar as the successful wintering of bees is concerned.

The first step in the packing process is to push the two colonies to-gether on a hive stand, creating a dead-air space below the colonies. The summer covers are removed and stored, but the inner covers with the hole open are left in place; the holes in the inner covers provide ventilation.

The Tarpaper Wrap

The most efficient method of packing bees is to wrap the colonies on hive stands that can accommodate two to four colonies. Hive stands built for two colonies are most common; when more than this number of colonies is packed in a group, there will be some drifting after the bees are packed in the fall and before they are unpacked in the spring. However, this is not considered to be a serious problem by most beekeepers.

Only a single layer of paper is needed for the wrap. For packing two standard ten-frame two-story hives, a piece nine feet four inches long will overlap five or six inches when wrapped around the two hives. No packing or insulation material should be used under the tarpaper on the sides of the colonies. A material that absorbs moisture, preferably wheat straw, should be placed eight to ten inches high on top of the colony's inner covers. If the holes in the inner covers are left open, there will be a good escape passage for moisture-laden air.

At the point the wrapping paper overlaps, two nails can be used to hold the paper in place. Nails are the easiest material to use to hold the top corners of the pack down. One piece of string—usually heavy binder twine—is used to tie the paper in place.

The effectiveness of the pack is best judged in the spring. The colony should be dry at that time. If any part of the in-

Colonies of bees wrapped with black building paper for winter. Steps in packing are shown, right to left.

Two two-super colonies packed for winter.

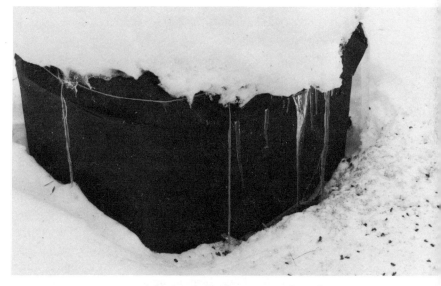

One virtue of wrapping colonies in black paper is that the snow melts first around the colony so that the bees may fly to void feces, as from this hive, in midwinter; the dark spots on the snow are feces, which bees will not void in the hive under normal conditions. The few dead bees on the snow are older bees that failed to return to their hive. The loss of some bees on the snow in winter is to be expected in the northern states.

terior is wet, then the pack is too tight. In more humid areas it may be advisable to put nail holes in the side of the pack to allow moisture-laden air to escape.

Mice in Beehives

Mice can be very destructive of beekeeping equipment. Not only do they destroy much stored equipment each year, but in the fall mice may invade and overwinter in active colonies. After the winter cluster is formed the bees abandon a large portion of the hive. The bees in a winter cluster cannot move from the cluster to protect all the space in the hive. The mice move into the hive at that time and chew out an area in the comb and surround their nest with leaves and other insulating material. This nest lining also prevents bees from attacking and driving the mice out of the hive when the weather warms sufficiently for the bees to move. The mice apparently learn to remain inactive at the time when the colony is active. In the spring a female mouse may raise a litter of young before moving out of the hive. Because the mice destroy so much comb, they should be poisoned, but this is a job that must be repeated each fall. No other method of preventing mouse damage is really satisfactory. This statement is not meant to be arbitrary but is based on the author's long experience with bees. Not everyone agrees with this position.

It is advisable to place poison for mice in the winter pack, both on top of the inner cover, under the hive, and within the colony itself. A few weeks before the packing is done in the fall some poison may be put in buildings in the vicinity of the apiary to begin to kill the mice. Since mice have the habit of moving into buildings, etc., in the fall, the job is not an easy one. Unless the beekeeper takes precautions to kill mice, much comb and equipment may be damaged.

Poisons for mice may be purchased in most drug and farm

stores. Formulas for preparing mice poison are available from county agricultural agents. The poisons used vary greatly in their stability and the more persistent types, made with strychnine, do a better job of ridding hives of mice. The use of this chemical is restricted in certain states; county agents can best advise concerning local regulations and laws.

Some beekeepers place hardware cloth, cloth with two or three wires per inch, over colony entrances. These serve as mouse guards. Their use can be successful, but it is a risky practice and not so effective as poisoning mice in the fall. Some mice are able to crawl through hardware cloth with one-half-inch holes, and therefore wire cloth with three wires per inch works best. However, hardware cloth of any size may hinder the removal of dead bee bodies from within the colony by the bees. If hardware cloth is used to protect the entrance against mice, it is a good idea to give the bees an upper entrance by drilling a hole in the second super; in the event the primary entrance becomes plugged, the bees are still able to fly.

Reducing a colony entrance with a wooden cleat so there is a passageway only three-eighths of an inch high is also an effective way of keeping mice out of a colony; but mice have been known to chew their way through wooden entrance cleats. Skunks and other animals can remove a wire or wooden entrance guard in an attempt to obtain food; this can provide an entrance for mice. Some beekeepers who use entrance cleats to protect against mice go to the trouble of fastening them in place with small nails or staples.

VIII

Predators and Diseases

Crowding a great number of bees into a single colony means that communicable diseases can spread quickly and easily. This is true in the case of some diseases and it is remarkable that honeybees do not have more diseases than they do. To certain large animals a beehive represents a great wealth of food; thus, it is to be expected that predatory, meat-eating animals, such as bears, can also be a problem.

In the United States all known honeybee diseases attack during the feeding stages and the eggs and pupae in the colony are little affected. Researchers have found that honeybees have several ways of protecting themselves against certain diseases that might attack the larvae or adult bees. Among these protective systems hive cleanliness is perhaps the most obvious to the beekeeper. Bees remove any objects they can carry. Large objects or animals such as mice or small snakes that might possibly die within the hive are encased with propolis. Bees that die in the hive are not merely removed to the hive entrance, but are carried many feet from the hive before being dropped by the flying bees. A favorite classroom illustration of hive cleanliness is to place small pieces of grass or straw into the top of an observation hive and watch how long it takes the bees to move them to the entrance and how far they will carry the pieces from the hive before they drop them. They sometimes carry such an object fifty to one hundred feet from the hive. Thus, from the practical point of view of the spread of disease, most bees that die within the hive are carried some distance and are not left close to the hive where they may be a further source of infection.

In addition to hive cleanliness, two other factors are important in the preservation of stored honey. Only one type of microorganism can live in honey; these are the so-called osmophilic yeasts. The osmophilic yeasts do not grow if the honey is properly ripened and contains less than about 19 percent moisture; however, if the honey is diluted, they will grow and cause fermentation. The high osmotic pressure of honey kills any other microorganisms that might accidentally contaminate the honey. This is one of the reasons that honey is a very safe food as far as humans are concerned.

In addition to a high osmotic pressure, it was discovered in 1963 that honey contains an enzyme, glucose oxidase. It is believed this enzyme is added to honey by worker bees. This material, an enzyme, which is active as long as the honey is not heated, continually releases a small quantity of hydrogen peroxide, better known as peroxide, a common bleach. The quantity of hydrogen peroxide released increases as the honey is diluted. The amount of this material present in honey is not sufficient to be detrimental when man eats honey, but it does have an effect on any bacteria that enter the honey. It is presumed that the hydrogen peroxide is especially helpful in preventing the contamination of honey that is diluted and fed to the larvae in the hive. Honey is diluted with water before it is used for larval feeding and this is one of the reasons bees need large quantities of water. It has been known for a long while that honey has an active antibacterial property; this has been described in the literature as the "inhibine" effect, but is now known to be the effect of hydrogen peroxide. The term "inhibine'" was coined in Germany in the 1930's to describe one antibacterial effect of honey.

American Foulbrood

American foulbrood is a bacterial disease that infects and kills only honeybee larvae. The disease is widespread throughout the beekeeping world and is well known to beekeepers as the most difficult of all bee diseases to cope with. Experimentally, attempts have been made to infect other insects and animals with American foulbrood, but the disease is specific and does not affect other species, including man.

There are two reasons why American foulbrood is difficult to control. First, a dead larva becomes a sticky, ropy mass which is difficult for the bees to remove from a cell in the comb. Second, the bacteria may form spores that are inactive but may remain alive for fifty or more years, awaiting the proper conditions in which to grow again. Not infrequently, colonies have died, the equipment has been stored and the disease reappears in twenty or more years when the equipment is again used for beekeeping.

People not familiar with bee diseases often confuse American foulbrood, European foulbrood and sacbrood. All of these are diseases of the larva but American foulbrood has certain characteristics that differentiate it from the others. Larvae attacked by American foulbrood usually die late in the larval stage or early in the pupal stage. Thus, the larva lies flat in the bottom of the cell and not infrequently one can see the remnants of a leg or tongue sticking up perpendicular to the dead body. After a month or so the dead larvae dry up into what is called a "scale," but prior to that time and while the color is still brown or dark brown, the remnants of the larva will string or "rope" out if a stick or hive tool is stuck into the dead mass. Not infrequently the larvae die after their cells have been capped.

In this case one will note that after a month or so the cappings are sunken and perforated.

Because it is not always easy to detect American foulbrood and differentiate it from other larval diseases, it is safest for the beekeeper to consult with the local apiary inspector or to send a sample of comb with dead larvae or a piece of a dead larva on a toothpick to the U.S. Department of Agriculture for identification. Samples addressed to the Bee Disease Laboratory, Agricultural Research Service, U.S. Department of Agriculture, Beltsville, Maryland, are diagnosed for disease without charge. Most states have an apiary inspector and he, too, may be consulted if the beekeeper is not certain of the disease.

Generally the infection of American foulbrood averages one to two percent in most states each year. The disease is spread both by beekeepers and by bees themselves. Beekeepers who transfer combs of brood or honey without inspecting the colony or divide a colony without inspecting for disease are not infrequently responsible for spreading the disease throughout an apiary. Apparently, too, bees may carry the spores externally on their bodies or internally in honey in their honey stomachs. Not infrequently bees drift from one colony to another and a bee moving accidentally from one colony to an adjacent colony can spread the disease. Bee inspectors have noted that it is common for two adjacent colonies to have American foulbrood, one in an advanced stage and one in a less advanced stage. In such cases it is thought that the disease was spread by drifting.

Just as bees can spread the spores of American foulbrood externally on their bodies, so beekeepers can spread American foulbrood with their hive tools or other equipment. This is one of the reasons gloves are not recommended for beekeepers in the apiary. It is not uncommon for gloves to become unusually sticky with propolis, etc. and to be a medium for American foulbrood exchange.

It is thought that the transmission of American foulbrood

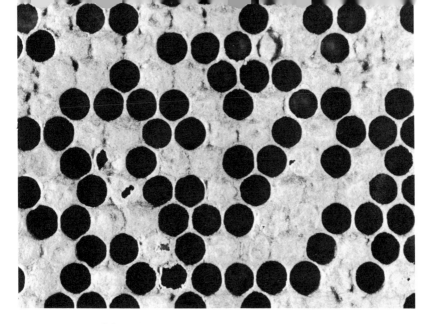

Capped brood dead in the pupal stage. Brood may die for many reasons; however, when the worker bees in the hive make perforations in the cappings such as these, the beekeeper should be suspicious of American foulbrood. This brood did die from American foulbrood.

When brood dies in the late larval or early pupal stage from American foulbrood, the remains first turn brown and form a stringy, ropy mass that can be pulled from the cell as illustrated.

could be cut if the colonies in an apiary were farther apart. It is recommended that different colors be used on hives, as we know that bees can see colors and use them for orientation. Also the presence of a tree or other object within an apiary will help in orientation and cut down on the number of bees drifting from one colony to another.

Treatment of American Foulbrood

The treatment for American foulbrood varies from state to state and area to area. The writer believes that the only effective way to deal with American foulbrood is to kill all infected colonies and to burn the equipment whenever the disease is found. It is not necessary to burn wood that is three-quarters of an inch or more thick; it can be scorched. However, the burning method is effective only where there is a good disease inspection program in force. This is not true in every state and some bee experts and beekeepers treat American foulbrood with drugs. Both sulfathiazole and Terramycin will kill American foulbrood in the vegetative stage. However, these drugs will not affect foulbrood in the spore stage. The problems in feeding drugs are many. If the drugs are fed to a colony at a time when nectar is readily available, not infrequently the drugs will be stored with other food and not be effectively used throughout the hive; it is extremely difficult to get a uniform dose to all larvae at the age when they are most susceptible to the microorganisms. It has been shown that a religious inspection program and burning system will effectively keep American foulbrood in check and hold colony losses to a level of 1 percent or less on an annual basis. One of the major reasons for discouraging drug use is that there is a very real danger of contaminating the honey. This could have an adverse effect on the beekeeping industry.

In some states it is recommended that drugs be used as a

preventive against American foulbrood infection; in such areas it is usually fed routinely, once in the spring and once in the fall. Near large cities this is often an effective method of controlling the disease. Inspectors have long noted that crowded city areas and areas near large dumps, where there might be exposed honey, are places where American foulbrood occurs more commonly. It is recommended that beekeepers consult with their neighbors, their state apiary inspector and state apiculturist or state entomologist to determine what is the most reasonable course of action in their area. The use of antibiotics should not be taken lightly nor should they be used by people not thoroughly familiar with bee behavior. Still, under certain conditions drugs can be useful in beekeeping.

Sulfathiazole is the favorite of the various drugs used for the control of American foulbrood. It is available from the manufacturers of bee supplies and is sold in tablet and powder form; sulfathiazole is most popularly mixed with sugar syrup, which is in turn fed to colonies. The precise quantities of drug and syrup to use are listed with the packaged material.

Early spring and late fall are the most effective times to feed drugs to bees as a preventive for American foulbrood. Early in the spring, when there is a large quantity of brood and a minimum of food, a gallon of medicated syrup fed to a colony is almost certain to be used immediately and effectively within the colony. At this time of the year, too, sufficient time will elapse between brood feeding and the honey flow that there is no danger of the drugs contaminating the honey. Again, in the fall, when colonies are adjusting winter stores, medicated syrup will be the last to be stored and the first to be used within the colony. The use of any drug requires considerable background knowledge. It is recommended that only those beekeepers who have consulted with the local apiary inspector or state apiculturist as to the precise situation in his state use such materials.

The Stress Diseases

Three diseases of honeybees should be considered as a unit: European foulbrood, sacbrood and nosema disease. Though their life cycles are vastly different and two affect larval honeybees and one, adult honeybees, they all manifest themselves under similar circumstances. When the colonies are under stress it is not uncommon to find all three diseases active in a colony at the same time, but it is not necessarily so that all should occur. The three diseases are more common in the spring and when colonies are rearing large quantities of brood.

Each of these three diseases can be exceedingly destructive and kill large numbers of bees within the colony. It is therefore important that the beekeeper do all that is possible to prevent the outbreak of these diseases within a colony and their spread within the apiary. The greatest stress placed on a honeybee colony is that of maintaining a brood-rearing temperature of about 92 degrees Fahrenheit. One factor that sets the honeybees aside from all other social insects is that they control brood-rearing temperatures and that they build a well-defined nest to facilitate their doing so. Nevertheless, at certain times of the year, especially in the spring, it is extremely difficult for the bees to maintain this temperature within the precise limits required. While other factors, such as poor food, a lack of fresh water, a shortage of pollen, etc., can also place stresses on honeybee colonies, it appears that maintaining a brood-rearing temperature is the most difficult chore the colony must undertake.

Effective control of the three stress diseases lies in the management of the colonies; this is especially true in the north. In the south one of the most common stresses placed on colonies is that of loss from insecticides, and the three distress diseases may manifest themselves under these circumstances. It is recommended that honeybee colonies be placed on hive stands four to eight inches high. If this is done, the bottom boards are not in contact with the damp ground, and grass growing around the colonies will not shade them so much. In most parts of the country honeybee colonies should be placed in full sunlight. In certain of the desert areas of the western states, especially Arizona, it is too warm to place colonies in direct sunlight so it is necessary to construct shade for them in these areas, but this is an exception. Additionally, it is helpful to supply bees with good quantities of fresh water. Honeybees use water to dilute the food they feed to larvae and also to cool their hives when they are too warm. It is felt that at least one of the diseases—nosema disease—can be spread by spores in contaminated water. Still another way to assist colonies in maintaining a proper brood-rearing temperature is to keep the colony entrances restricted in the early spring. How much a colony should be restricted will vary with the location and strength of the colony.

Stress diseases often develop in colonies that have lost large numbers of bees because of pesticides. It is not infrequent that beekeepers who move bees into orchards or other places for pollination where there is pesticide contamination find the diseases manifesting themselves a week or several weeks later. There is no good protection against this kind of stress other than to make sure that the colonies are otherwise strong and healthy.

European Foulbrood

European foulbrood is a bacterial disease that affects only larval honeybees. Unlike American foulbrood, the European foulbrood organism does not form a spore, so it does not pose the great threat that American foulbrood does. Larvae killed by European foulbrood usually die at a younger larval stage than those killed by American foulbrood. Larvae killed by European foulbrood are usually curled or twisted in their cells whereas those killed by American foulbrood lie perfectly flat. Also, larvae killed by European foulbrood do not have the sticky, ropy consistency of those killed by American foulbrood.

In certain parts of the country it is recommended that Terramycin be used for the treatment of European foulbrood. It appears that this drug is an effective control measure, but it should be used only when it is impossible to eliminate or reduce the stress that brought about the disease.

Certain races of bees are more susceptible to European foulbrood than others. Prior to about 1880 the black or German race of honeybees of Europe was the most popular one used in the United States. However, this was replaced by the Italian race of bees. Even today, when severe infections of European foulbrood manifest themselves, it is recommended that the colonies be requeened with good, young Italian queens.

Sacbrood

Sacbrood is a virus disease that affects only honeybee larvae. The disease is so called because the dead larvae may be lifted from their cells and appear saclike. There are no drugs or medications that have any effect on sacbrood and the only protection the beekeeper has are the protective measures that are taken against the other stress diseases.

Nosema Disease

Nosema disease is caused by a microorganism that affects only adult honeybees. The microorganisms invade the cells in the gut of the bee and destroy them, thus upsetting normal digestive and physiological processes. It has been shown that nosema disease may be found in almost every apiary in almost every part of the country. However, it is quite definite that it is usually only in the early spring and when colonies are under stress that the disease manifests itself. The chief symptom is the presence of a large number of weak, crawling bees in front of the colony entrance. Nosema disease may be accompanied by dysentery, which may be brought about either directly by nosema disease or by poor food within the colony. Fumagillin has proved to be fairly effective against nosema disease, but its cost usually prohibits its use by most beekeepers.

A chief complaint from package-bee users and queen users in the northern states and Canada is that bees are shipped from the southern states infected with nosema. While this is probably true, the same nosema disease causes little or no difficulty in the southern states. Because of the cooler temperatures in the north, a stress is placed on the bees and the disease can kill or shorten the lives of a large number of bees in the package shipped north. Drug feeding is recommended and is probably effective in Canada; it is questionable whether it is needed in most of the northern United States, though even there it may be helpful. Certain package-bee and queen producers feed the drugs in the southern states. While feeding drugs in the south may not show any visible improvement in the colonies there, it undoubtedly reduces the incidence of infection in the bees and increases the chances for successful development of the package in the North.

Other Microbial Diseases

While the four honeybee diseases disscused above are the major ones beekeepers will encounter in the United States, there are several other, lesser-known maladies that affect larval and adult honeybees. These include amoeba, chalk brood, parafoulbrood and paralysis, to mention only a few one can find listed in the literature.

Generally speaking, these are diseases that cause little loss and for which no special cure is necessary. In part, it is expected that these, too, are diseases that manifest themselves more under stress, and the routine practices recommended above will do much to eliminate them from the apiary.

Chilled Brood

Not infrequently, in the early spring, when the weather is favorable, honeybees will attempt to rear more brood than they can keep warm during a period of adversity. Beekeepers have noted that, if a week of inclement weather follows a week of unusually warm weather in the months of April and May, in the northern states, some larvae and/or pupae on the outer edges of the brood nest are chilled and die. If larvae are killed by excessive cold, they are usually removed by house bees within twenty-four hours and are seldom noticed by the bee-keeper. Contrariwise, cells containing dead pupae are not un-capped immediately nor are the pupae removed; the removal process may take several weeks.

In the United States there are no adult bee diseases that kill in the late pupal stage. Thus, when one finds dead pupae within a colony, especially in the early spring, it is almost certain that they died of chilling.

There are no special precautions to be taken to prevent chilled brood except to keep the entrance cleats in place as long as the weather is cold and the colony population is not so great that they must be removed. Chilled brood can occur in colonies of large populations as well as of small populations, though it is more likely to occur in the latter. Care should be exercised not to divide colonies too early in the spring to make new increase (new colonies).

Mice

In the fall months throughout the northern United States it is common for mice to move from their field nests into buildings and other areas where they can build better-protected nests for winter. Mice often nest in stored combs and in hives containing bees. It is an interesting fact that mice can successfully build a nest within a honeybee colony with a strong population and survive the winter without difficulty and without being stung by the bees. However, the mice are protected by a thick layer of nesting material and no doubt move in and out of the colony entrance at a time when the bees are inactive. Mice should be eliminated from the vicinity of the apiary as well as possible. (See "Mice in Beehives," page 103.)

Bears

In the mountainous and heavily wooded areas in many parts of the United States bears are common and will sometimes attack beehives. Not all bears are a problem to apiarists. Contrary to a popular opinion, bears are more interested in eating brood than they are in eating honey; however, beekeepers talk very little about this point as they do not care to destroy the popular image of a large brown bear enjoying a meal of fresh honey.

There is no perfect protection against damage by bears. Electric fences have been devised to keep bears out of apiaries, but their cost is too exorbitant for most persons. The return a commercial beekeeper can expect for investing in a bear fence is not sufficient to warrant its construction. Bears are wary animals and are less inclined to attack bee hives in the vicinity of a house or barn; however, this is not always an effective measure for keeping bears out of an apiary.

The only easy and effective ways to rid an apiary of a bear is to poison it or to trap it. Both methods are illegal in most states, but in most states it is legal to shoot bears. It is extremely difficult to do so, since bears usually become quickly aware of the presence of people. In certain states conservation

departments trap bears in large, humane-type traps and move
the bears to areas where they presumably will not be a problem.
However, bears have been known to walk long distances to
reach a good source of food.

At least four states—Pennsylvania, Vermont, New Hampshire
and Minnesota—compensate beekeepers for losses as a result of
bear damage. The reasoning behind these laws is that bears are
protected for hunting purposes and that the state should be
liable for their depredations. Not all conservation departments
take this view.

People who have suffered damage from bears report that the
bears will pick up and carry colonies as much as two hundred
yards from an apiary before dashing them apart on the ground.
It is also generally agreed that when a bear begins to work in
an apiary, it takes one to three colonies a night and continues
to do so until the whole apiary is destroyed. While the bears
may be stung considerably, this does not seem to deter them.
Despite their great liking for brood, there is no doubt that the
bears consume honey and also eat some adult bees.

Skunks

Common skunks, like bears, are primarily meat eaters and
live largely on insects and small animals which they are able
to catch. Skunks are a serious problem for beekeepers in certain areas of the country. The usual procedure is for the skunk
to feed in the evening, scratching on the hive entrance so as
to disturb the bees and cause them to fly out or to move to the
hive entrance to determine the source of the annoyance. The
skunks swat and attempt to kill the bees as they crawl through
the entrance hole of their hive and then eat them.

Skunks have been dissected, and it was discovered that their
tongues, mouths, throats and stomachs all had bee stings, but
apparently their liking for the insect was sufficient that this did

not deter them. People who have observed skunks feeding have noted that they roll, twist and turn on the ground in front of the hive as they are stung by other bees coming from the hive. Evidence that skunks have been feeding on colonies of honeybees can be seen by observing the entrance and the grass immediately in front of it. The entrance of the colony is usually covered with mud and the grass and sod are torn up in front of the beehive.

The only effective ways to prevent skunk damage is to kill the skunks with poison or to move the bees. Poisoning skunks is illegal in most states as they are legally protected as furbearing animals. If the skunks are properly poisoned, usually with strychnine, an objectionable odor will not be left in the vicinity. Skunks should not be trapped in an apiary, since a skunk in a trap releases considerable quantities of skunk odor and it is difficult to rid the area of it for some time. Fencing an apiary or elevating the colonies may prevent skunk damage, but these are not very practical ways of dealing with the problem.

Squirrels

Red squirrels can be a serious menace in bee equipment stored in buildings, especially outbuildings that are not inspected very frequently. The squirrels find a building a dry place in which to nest and during the winter months they can chew and destroy many combs, presumably to obtain honey and pollen for food. Most beekeepers leave poison wheat or bait in their buildings to kill any rats or mice; however, squirrels usually do not eat such bait. Since it is very difficult to shoot or trap squirrels in a building, it is best to poison them. Squirrels will eat walnuts containing a small amount of strychnine, but the use of strychnine is dangerous where there are children.

Wax Moths

There are several species of beetles and moths that infest and destroy honey comb. Of these the common wax or bee moth, *Galleria mellonella,* is the most common and most destructive pest. Wax moths will infest stored equipment and are sometimes found in portions of active honeybee colonies where the comb is not covered by bees. Wax moths are more destructive in the southern states than in the northern states. In the southern states it is felt that wax moths are a valuable adjunct to the American foulbrood inspection service. Wax moths can very soon destroy a colony that dies from disease and in doing so probably render a service to the beekeeping industry.

Several materials can be used to control wax moths effectively: paradichlorobenzene, methyl bromide, carbon disulfide, sulfur, and ethylene dibromide. Not too many years ago calcium cyanide was widely used by beekeepers to fumigate stored combs. Calcium cyanide is highly toxic and there are many records of beekeepers dying from this material through over-exposure as they were fumigating combs. It is recommended that beekeepers use any one of the five materials noted above rather than cyanide.

Paradichlorobenzene is probably the most widely used fumigant, among beekeepers, for wax-moth control in stored supers. This is the same substance that is popular with housewives for protecting stored clothes against clothes-moth damage. Paradichlorobenzene has a low mammalian toxicity, it vaporizes slowly over a long period of time, is not toxic to honey bees, and is easy to use. Commercial beekeepers who must protect hundreds or thousands of supers of honey combs construct windowless rooms, sealed as tightly as possible, which they use for fumigation and as storage chambers. Beekeepers who wish to protect only a few supers against the ravages of the wax moth fumigate the supers while they are stacked in piles. The most effective way to fumigate stacked supers is to place a newspaper between every third or fourth super in the stack and to place a small amount of paradichlorobenzene on the paper. The fumigant is effective as long as a small quantity is present.

Paradichlorobenzene should never be allowed to come into direct contact with stored combs or beeswax that is to be rendered, as the crystals may dissolve into the wax and spoil it. It is preferable that treated supers should be exposed to fresh air for several days before they are placed on colonies. However, this is not always done and there do not appear to be any adverse effects when the supers are not aired.

Wax-moth eggs, larvae, pupae and adults are killed by freezing temperatures. For this reason wax moths are seldom a problem for northern beekeepers until August or September; then it is especially important to check stored equipment and to give it adequate protection. The wax moths responsible for late-summer infestations may be the progeny of moths that migrate north, probably from states at about the latitude of the Carolinas, or they may come from infestations that survive the winter in combs stored in warm cellars or partially heated garages in the north. One method of combating the wax moth is to store all comb and beeswax refuse in unheated buildings in the winter where all stages of the wax moths will be killed. Storage in such buildings may expose the comb to rats, mice and squirrels so precautions must be taken against these animals, too.

At the time of this writing none of the fumigants above is approved by the Pure Food and Drug Administration for the fumigation of comb honey. This places the comb-honey producer at a great disadvantage as it has always been recommended that comb honey be fumigated prior to being placed on the grocer's shelf. Wax-moth adults lay their eggs externally on supers when they cannot gain access to the hive because of being attacked by worker bees. Upon hatching, the first instar wax-moth larvae are capable of traveling many inches to obtain food and if eggs have been laid on the exterior of a comb-honey super, it is possible the sections of comb honey may be infested by the time they reach the market.

Beekeepers who produce a small number of comb honey sections find they can best protect them against wax-moth damage by placing them in a freezer for a few to twenty-four hours. Freezing the sections, prior to marketing them or giving them to friends, will kill any wax-moth eggs or small, barely visible larvae which might be on the wood or wax surface. Contrary to popular opinion, freezing the sections does not have an adverse effect on them nor does it cause them to weep (leak honey from the cells). In fact, comb honey sections may be stored in a freezer for many months; under these conditions the honey will not crystallize and the sections will retain their flavor. The greatest danger in freezing comb honey sections is that the cold, frozen sections fracture easily as they are being removed from the freezer.

In the laboratory it has been shown that all stages of the wax moth are killed by exposure to 120 degrees Fahrenheit for twenty-four hours; the eggs are killed by exposure to this temperature for even a shorter period of time. However, no one has tested the use of heat as a practical control measure; since comb honey is such a delicate product, and beeswax melts at 148 degrees Fahrenheit, the heat treatment is obviously a risky practice. Under the present federal regulations it behooves beekeepers to be more careful than ever to store unused equipment under circumstances where the moths will be killed by freezing temperatures in the winter.

Mites

No honeybees have been imported into the United States

since 1923 with the exception of some live eggs and larvae brought in for research purposes in recent years. The major reason for instituting the law in 1923 was to keep out of the United States certain bee diseases, especially acarine, caused by a mite, *Acarapis woodi,* which is prevalent in Europe and certain other parts of the world. While we have little data as to how potentially harmful *Acarapis woodi* might be in the United States, it is generally agreed that we do not need to import other races of bees into this country, and there is no need to endanger the beekeeping industry by doing so. In the years that honeybees were imported into the United States a great variety of races and strains of bees from various parts of the world were brought here and most of these can still be found in varying conditions today. Generally speaking, most commercial apiarists consider that the management of honeybee colonies is a far more important consideration than the race or strain used. Within the past few years it has been discovered that there are two species of mites, *Varroa jacobsoni* and *Tropilaelops clareae,* which infest at least two of the three species of honeybees found in Asia. It is the opinion of the writer, after two research trips to Asia, that the reason for the failure of *Apis mellifera,* the European honeybee, in Asia is the presence of these mites, which transfer readily to imported bees. It is recorded in various bee journals that several hundred attempts have been made to introduce our honeybee into India and other parts of Asia during the past several centuries; these importations have always failed and will probably continue to fail until some method is found of coping with the mite.

Unfortunately some researchers have imported one species of honeybee, *Apis indica,* into Europe on several occasions. It is also believed that one importation of *Apis indica* was made into the United States several years ago. So far as has been determined the mites have not been found in either of these areas, though an extensive search should be made for them in the areas of importation. These importations occurred prior to man's discovery of the presence and importance of the honeybee mites in Asia. These importations indicate that great care should be exercised in moving sundry plants and animals from one continent to another.

Pesticides

Pesticides are a much talked-about problem for beekeepers in all countries where bees are kept. American technology has changed the agricultural picture so that today only about 7 percent of our population produces food for all the rest of us; pesticides are an intricate part of this new agriculture. An intensified agricultural system would not be possible without the chemicals to control all the insects and diseases that affect so many crops.

While growers may take several precautions to protect honeybees against pesticide losses, they seldom do so because of the pressure of time, etc. Growers of agricultural crops are usually behind in their work and concerned about the progress of their crops and as a result use pesticides at times when they can cause inconvenience to others. There are several things that growers can do to assist the beekeeper, including selection of the proper pesticide, use of the right dosage, use of granular materials when possible and selection of a time of day for application when honeybees are not active. In cases where beekeepers are renting bees for pollination they can insist that certain of these precautions be taken.

The beekeeper's best protection against pesticide loss is to become as knowledgeable as possible about the agricultural practices in his area. Most farmers and agricultural agents are sympathetic to the use of bees and understand their value in agriculture today. In an intensified agricultural system certain crops could not be grown without bees. Often a little time spent with growers in the winter and a gift of a few jars of honey will do much to eliminate losses in the future. When losses of honeybees from pesticides are observed, it is important to inform the state entomologist, state apiary inspector, state apiculturist and especially the local agricultural agent. This alerts several people to the problem and the dangers concomitant with the use of certain pesticides. The local agricultural agent is the man who makes recommendations concerning which pesticides should be used; if he is made aware of local problems, he will be in a better position to make recommendations that will protect the bees in his area.

The Toxicity of Pesticides

Insecticides do not have the same effect on all insects. DDT, for example, which has a serious effect on a wide range of insects and other animals, has only a slight effect on honeybees. In fact, it is difficult to kill a honeybee colony even with a direct application of DDT. Carbaryl (Sevin), on the other hand, a relatively new insecticide that is favored because of its rapid breakdown and the fact that it does not accumulate in nontarget organisms as does DDT, has caused the loss and/or weakening of thousands of colonies of bees throughout the United States. In fact, as far as beekeepers are concerned, it poses more problems than any other insecticide.

When insecticides are prepared for the market, they are mixed with a variety of other materials such as oils, talc, spreading and/or wetting agents, etc. These combinations are called "formulations." Formulations vary in their toxicity even though they may contain the same amount of insecticide. Thus, in writing about materials it is difficult to say what the toxicity of a particular material might be.

Researchers in California and the state of Washington, where insecticides are used in great quantity, have prepared lists showing the toxicity to honeybees of various pesticides.

These lists are revised annually and are helpful in determining whether or not a material will cause trouble in a new area. Copies of these lists can be obtained by writing the state apiculturists in the respective states.

Protecting Colonies Against Pesticide Loss

Worker bees usually forage within about one-half mile of their hive; however, when food is not available nearby, foraging bees may fly for two or three miles in search of it. Most honeybee-pesticide losses take place as a result of spraying within half a mile of the hive or apiary; however, losses at greater distances have been recorded. The author once examined an apiary where half of the approximately fifty colonies showed a loss as a result of spraying two miles away. In this case some of the colonies obviously had a number of foragers in the spray area while the field forces from other colonies were foraging in other directions.

Clearly, a beekeeper cannot be responsible for knowing about all of the agricultural or forestry projects in his area and often he suffers bee losses without ever being able to determine the cause. Where a beekeeper is aware of impending spraying there are only a few choices open to him. More and more beekeepers are inclined to do nothing if the spraying is taking place more than a mile from their apiary and if there is a reasonable quantity of forage available to the bees.

Some beekeepers have successfully covered their colonies with wet burlap so as to confine the bees during a spray program. But this is a risky, expensive and time-consuming procedure that few people would recommend. The burlap must be wetted several times a day, especially on a warm, dry day. Since most pesticides remain toxic for several days, the bees are usually covered for at least two days. Such confinement upsets the routine of the colony. More important, normal ventilation is interfered with and colonies often overheat and die as a result.

Moving colonies away from a spray area for one or two weeks is the best protection when a beekeeper is given sufficient advance warning of a spray program. This requires that a spray-free area be available for the colonies. One state, New York, provides compensation for beekeepers who are forced to move colonies as a result of a state or federal spray program.

Compensation

In 1971 the United States Congress enacted new legislation to compensate beekeepers who lose bees as a result of the application of any federally registered pesticide. The actual legislation is listed under Section 804 of the Agricultural Act of 1970. The program is retroactive and covers losses dating back to January 1, 1967. The amount paid per colony varies, depending upon the extent of the beekeeper's loss.

This legislation climaxes a long argument over grower-beekeeper responsibilities and the value of the honeybee in the economy. Some have argued that the beekeeper should be responsible for where his bees fly; the counterargument is that the honeybee is an important part of the economy both as far as practical agriculture is concerned and also for the pollination of fruit, nut and seed crops for wildlife; wildflower pollination is also a consideration.

The indemnification program is under the jurisdiction of the Agricultural Stabilization and Conservation Service of the U.S. Department of Agriculture. A handbook on the beekeeper indemnity payment program is available as well as forms for making application for payment. Copies may be obtained from ASCS, which has an office in most counties in the country. Also county agricultural agents can provide further information.

IX

The Queen

The queen is the heart of the colony. As a result of talking to commercial beekeepers, it is clear that they have great difficulty in obtaining and keeping good queens. This is especially true with those who produce honey in the northern states and Canada, where their colonies must be in peak condition to gather the crop during the short season. Not infrequently too, beekeepers say that locally bred queens are better than those they can buy from southern queen breeders. This latter statement is sometimes, but not always, true.

Queens can vary. How they will vary depends upon the food and treatment they received during the larval stage. While beekeepers may observe differences in the size of queens, it is usually the number of ovarioles that researchers use to determine queen development. The object in rearing queens is to rear them under optimum conditions. We know that it takes a good supply of pollen and a good supply of honey to rear a good queen. Unfortunately the person who buys queens does not always know when they were reared or whether they were reared under the proper conditions.

The role of the queen in the honeybee colony is twofold. She is responsible for laying the eggs in the colony. At the same time, she produces in glands in her body various chemical secretions that control the social order in the beehive. Our knowledge of the nature of secretions produced by queens is still meager and is largely a result of research undertaken during the past decade. However, we have sufficient information to know that it is through queen secretions that social order in the honeybee colony is controlled.

A honeybee colony contains a single queen. Rarely will one find two queens and when a two-queen condition exists, one is the mother and the other the daughter. No one has ever reported sister queens living together in a colony. When two queens are present, the older of the two has lost most of her queenly abilities, and while she may lay an occasional egg, she is deficient in the chemical secretions that control social order. One researcher told me he had found a colony with three queens all of which were marked and which he knew to be grandmother, mother and daughter; however, this is obviously extremely rare.

A queen has a long life-span and can live from two to five years, but rarely longer. On a commercial basis queens are usually replaced every year or two. The fact that a queen lives so long suggests that it is the hard work of flight activity and foraging of worker bees that is responsible for their short life.

Our Understanding of Queen Substance

Queen substance is a material produced in the mandibular glands of the queen honeybee. She distributes the secretion over her body, presumably with her legs. Queen substance and related materials are attractive to worker bees and it is for this reason that they surround and lick their queen. Some researchers have said that the attention given to the queen by the workers is such that one might very well say the workers are addicted to certain queen secretions.

The principal ingredient in queen substance is a relatively simple chemical compound containing only carbon, hydrogen and oxygen. The chemist calls the material a "ten-carbon fatty acid." It contains ten carbon atoms, sixteen hydrogen atoms and three oxygen atoms.

In the biology of the honeybee, queen substance plays several roles; three of these are reasonably well known. The fact that one material may do so much is remarkable but indicates that nature strives to conserve substances and energy. The three roles played by queen substance are quite distinct and there is no confusion about these matters among the members of the colony.

Queen substance is responsible for queen recognition by the workers. We know that, if we remove a queen honeybee from a colony, the bees will attempt to rear a new queen. They become aware of the queen's absence within hours and within twenty-four hours one can find new queen cells in the colony. In the laboratory it has been shown that the synthetically made queen substance will inhibit queen rearing. However, we have not been able to translate this research into anything practical. The problem lies in the fact that we do not understand precisely how queen substance is distributed within the colony and we have not been able to duplicate the natural distribution of this material on a large scale or within a large colony.

The second role of queen substance is to inhibit ovary development in worker bees. Again, beekeepers have known for many years that, if a queen is removed from a colony and the colony prevented from rearing a new queen, the ovaries of certain workers within the colony will develop and these workers will lay eggs. Laying-worker colonies pose a serious problem for the beekeeper because the process is irreversible. Worker bees become laying workers when they do not receive sufficient quantities of queen substance and perhaps other materials from the queen. Even after this balance is restored, they continue to be laying workers. There is no way of successfully saving the bees in a laying-worker colony.

Queen substance is also the honeybee sex attractant, its third role. Virgin queens, upon emergence, produce no queen substance. It is not until several days after they have emerged from their cells that the material is produced. It is interesting that within the honeybee colony drones pay no attention to queens. It is assumed that the quantity of the sex attractant is so great that the sensory receptors of the drones are overwhelmed. Men are familiar with circumstances when they have worked very close to an odor, objectionable or otherwise, to which they became so accustomed they could no longer detect it. Outside of the hive drones readily detect virgin queens. It can be shown that male honeybees will pursue and attempt to mate with wads of cotton, pieces of wood or any other material anointed with the sex attractant.

Interestingly, worker honeybees that are very attentive to their queen within the colony are antagonistic toward her outside of the colony. While conducting experiments on the sex attractant and mating, several researchers have reported that queen bees they were holding were attacked and even stung by worker bees outside of the hive. However, it was found that nature had found a way to circumvent this problem. Worker honeybees normally fly within about eight feet of the ground. They will fly higher if they are foraging on a tree or if it is necessary for them to fly over a hedgerow or section of woods to get to the area where they are foraging. However, under normal circumstances, their flight lanes are close to the ground. On the other hand, drones and queens seldom fly within twenty feet of the ground. They may fly at heights of thirty, forty and even sixty feet above the ground and mating takes place, for the most part, at these heights. When there is a strong wind queen and drone flight may be closer to the ground, but likewise, such a strong wind would likely lower worker flight, too. Sometimes one can observe a mating queen and drone fall to the ground during the mating process, but they remain there only briefly.

Queen substance is called a "pheromone." A pheromone is a chemical substance secreted by one individual which has an effect on another of the same species. A pheromone is a chemical messenger, a material that conveys a message to another animal of the same kind. It has been known for a long time that odors play an important role in honeybee colonies, but it is only recently that the biology and the chemistry of the problem have been understood.

Queen substance is not the only pheromone present in the honeybee colony. One of the best-known pheromones is the alarm odor. This is a chemical substance secreted from the vicinity of the sting and contains seven carbon atoms, fourteen hydrogen atoms and two oxygen atoms. Its chemical name is isopentyl acetate. The material has an odor not unlike that of banana oil. If one presses with his finger the thorax of a worker bee at the entrance of a colony, he can see that the bee protrudes her sting and soon other bees rush out and inspect the worker that is giving alarm. They may also sting the individual holding the bee! In the process of protruding her sting the worker bee releases the alarm odor.

Other pheromones include scent-gland secretions and materials that help to orient swarms as they are moving through the air. It is now thought that there may be twenty to one hundred or more chemical substances produced by worker bees and queens that dictate social order within the colony. This elaborate chemical communication system is still not fully understood. In fact, it has been studied for only a few years and much research remains to be done before we can fully define a honeybee colony in terms of its pheromones.

Mating

A colony that is rearing a new queen may produce one or

many queen cells. It is generally stated that a colony that is superseding their queen will produce one to six queen cells, while a colony that is producing queens for swarming may make four to twenty queen cells. While this very general statement appears correct, it is known that there are great differences in this regard. Some colonies produce far more queen cells than others under what appear to be similar circumstances.

Upon emergence, the first virgin queen kills the other queens, either before they emerge from their cells or in actual combat with them. The only time when this is not true is when a virgin queen leaves the hive with a swarm. During inclement weather workers may confine emerging virgins in their cells and prevent them from emerging and fighting. Virgin queens may accompany a primary (first) swarm if the old queen is lost; some colonies cast off secondary and tertiary swarms, taking virgin queens with them. Secondary and tertiary swarms, etc., are more common in certain races of bees than others.

Upon emergence from her cell a virgin queen honeybee is unattractive to workers or males within the colony. After about twenty-four hours some worker bees may be slightly attracted to the queen, but it is not until after she has made mating flights and begun to lay eggs that it can be truly said that workers are attracted to her and surround and groom her.

Virgin queens usually take their first flight five or six days after emergence. They, like worker bees, may use the first flight for orientation only and mating may or may not take place. Researchers have found that about 40 percent of virgin queens mate on the first flight. Flights last from a few to thirty minutes. No one has ever seen a queen bee alight or visit a flower in the field; queens and drones feed only within the colony. For this reason the flight time of queens and drones is restricted by the quantity of food they carry with them. The number of mating flights a queen makes is dictated by the quantity of sperm packed in her spermatheca on successive matings. A normal queen mates until she has slightly over five million sperm, which will last her the remainder of her life. If drones are few in number in the area, the number of mating flights will be greatly increased. Queens start to lay eggs eight to fifteen days after emergence from their cells. Rarely will a queen mate as long as thirty days after emergence; it is generally stated that a queen cannot mate normally after about twenty-five days of age. Once she has received a full complement of sperm, the queen does not leave the hive again unless it is to accompany a swarm.

Prior to 1951, it was thought that queens mated once or, rarely, twice; this is stated in the textbooks. In 1951 a Russian worker suggested that queens mated more than once. Mr. S. Taber III, of the U.S. Department of Agriculture Bee Laboratory, confirmed this in 1954, stating that the average queen mated seven to ten times. Mating may take place within the immediate vicinity of the apiary, but most generally queens fly to areas called "drone congregation areas." These are locations where the males congregate and apparently wait for the mating queens. While queens may fly a few miles to mate, probably most successful matings take place within half a mile of the apiary.

Natural mating in honeybees was not observed closely or accurately described until the early 1960's. Queens and drones mate at heights of twenty to possibly one hundred feet. When virgin queens, or the synthetic queen substance, are elevated on towers or with a helium-filled balloon, the activities of the males may be observed. While tethered queens mate only rarely, this has been seen often enough to describe the process.

In 1927 Dr. L. R. Watson discovered a method of instrumentally inseminating queen honeybees. This technique has been widely used as a research tool and has been improved upon by several researchers. While the technique has been known for a long while and used by many people, it has had little impact on commercial honey production. Although so-called hybrid queens are available, their sale does not dominate the field and only naturally mated queens, even if they are derived from instrumentally inseminated mothers, are used for commercial honey production.

From what has been stated, it is evident that bee breeding is still in a primitive state. We still know too little about the mating processes in honeybees and we have not been able to control breeding lines in the way that we would like. Many attempts have been made to store honeybee semen, in the same way in which sperm is stored from other animals, but these techniques have not proven successful over long periods of time. Researchers have studied mating honeybees in confinement and have attempted to mate tethered queen honeybees but without success. Research in this area and the development of better strains of honeybees for pollination and honey production would be most useful.

The Value of Our Knowledge of Pheromones

Our knowledge of the chemical substances exchanged by honeybees within the colony has been of little practical value to date. However, it has served to explain certain phenomena that have been long known to occur within the hive. For demonstration purposes we can show how certain of these factors work. For example, we understand better how swarms are able to follow their queen. We know why bees surround and lick the queen. We understand that, when there are two queens in a colony, one of them is no longer producing the chemical substances that makes her queenlike.

This new knowledge emphasizes the importance of having a good queen in the colony. Whereas to date we have been concerned about the size of queens and ovarial development, we now understand that glandular development is equally important.

In bee-breeding programs the emphasis has always been on honey production. Colonies that produce large quantities of honey are usually those from which queens are taken for breeding purposes. However, honey flows are relatively short and may be misleading as indicators of a queen's capabilities over long periods of time. If we are able eventually to assess the production of certain chemicals by queens within a colony, we may be able to breed better bees.

Reserve Queens

Even annual requeening of honeybee colonies does not guarantee that some queens will not fail during the course of the active season. It is recommended, and is usually a commercial practice, to keep a small number of queenright (queen present) nucleus colonies in each apiary for requeening purposes. Most beekeepers agree that having a number of reserve queens equal to about ten percent of the colonies in the apiary is not too many. Reserve queen colonies need not be large; they should contain between one and two pounds of bees and a young queen. The small colony is usually kept in a standard ten-frame hive.

While there are a variety of ways of requeening colonies that either have lost their queen or have failing queens, uniting the two colonies using the newspaper method appears to be the most popular and successful. This method involves placing one colony on top of the other so that they are separated only by a single sheet of newspaper. Usually five or six small slits, each two to three inches long, are made in the newspaper. When two colonies are placed together in this manner, the bees slowly eat away the newspaper until it is completely gone, usually in twenty-four to forty-eight hours. Because the bees from the two colonies are forced to mingle slowly, there is a mixing of colony odor and very little fighting of the worker bees. Of all the known methods of requeening, this one is probably the most satisfactory. When one colony is weaker than the other, the weaker of the two units is placed on top. Usually, too, if one is requeening with a small nucleus colony, this will mean that the queen is on top of the two combined hives; however, usually after a couple of weeks it will be found that she is forced down into the lower parts of the hive.

Small nucleus colonies with reserve queens should be prepared in May or June by making a new colony from an old one using one frame with some brood and shaking into the hive bees from two or three other frames. It is good to give the colony a comb of honey and also a comb of pollen. If the colony is not used for requeening purposes, it can be divided in July or August to make a second reserve queenright colony. Such colonies may be needed up to the time that honeybee colonies are packed for winter in the fall. If the reserve queen colonies are not needed by that time and are still in single supers, two of them can be combined to make an overwintering colony. Usually by fall they have enough bees and honey that this can be done with ease.

X

Special Practices

The production of honey is the primary reason that most people become interested in beekeeping. However, honey production involves heavy lifting and the management of large, populous colonies. In some areas the quality of the honey is not so good as it is in others, thus discouraging certain amateur beekeepers. At the same time, the lack of a good honey market or the location of an unusually good market may stimulate interest in honey production or in other aspects of apiculture.

There are a number of auxiliary pursuits for both hobbyists and commercial beekeepers. Certain of these are dependent upon location and local need. Books and papers have been written about some of the subjects discussed below. An invaluable source of information for those who care to pursue a particular subject further are the local county agricultural agents and the apiculturists at state colleges.

Queen Rearing

A fertilized egg laid by a queen honeybee develops into a worker honeybee or a queen, depending upon the food received during larval life. When this fact was first learned in the 1800's, beekeepers attempted to rear queens. For many years the only technique was to divide a colony and to allow the queenless half to rear a new queen, a method still used today in parts of the world where queens are not readily available.

Practical methods of rearing queens were not discovered until 1883, when Henry Alley of Massachusetts found that he could introduce pieces of comb cut into strips, with eggs in alternating cells, which would be accepted and enlarged by colonies of queenless bees to produce queen cells. A few years later G. M. Doolittle found a method of making artificial queen cups and of grafting (moving) young worker larvae into these. By the late 1800's the business of queen rearing was well developed and large numbers of queen bees were being shipped by mail. The basic techniques of rearing queens developed in the late 1800's are still widely used today.

Northern honey producers frequently complain that most queens are produced in the Gulf states or California and that these queens are inferior to those produced in the northern states. While there is no proof that this is true, it may be so. People who rear queens are well aware of the fact that better queens are produced when there is an abundant supply of both pollen and nectar. While queen larvae are reared on a glandular secretion from worker bees—royal jelly—workers cannot produce this secretion without an abundance of food. Studies have shown that queens produced from well-fed larvae are larger and have more ovarioles per ovary. It is presumed that queens with more ovarioles produce more eggs. Queen breeders usually keep sugar-syrup feeders in queen-rearing colonies as a safeguard. But, more important, queen-rearing areas are selected because they have an abundant supply of both pollen- and nectar-producing plants during the queen-rearing season, which is February through May and, to a lesser extent, June, July and August.

Queen mating nuclei in Florida.

While it is true that there is a shorter season for queen rearing in the north, it is difficult to believe that this is the whole reason that few queens are reared in the northern states. It is entirely possible that the whole matter revolves around a lack of knowledge in the north and the fact that the honey flow is more intense in this area, making honey production more profitable.

Rearing queens involves transferring (grafting) about twenty-four-hour-old worker larvae from their cells to slightly larger, man-made queen cups. The queen cups are placed in colonies of queenless bees, which usually accept and feed the larvae lavishly. After twenty-four to forty-eight hours in the so-called starting colonies, the newly grafted larvae are transferred to what are called "finishing colonies." While starting colonies have no free flight, finishing colonies are queenless colonies from which the bees are allow to fly. The larvae in queen cells continue to be fed in the finishing colonies and the cells are capped with beeswax on the fifth day of larval life. One day before the pupae are due to emerge from their cells, which is fifteen days after the egg has been laid by the queen, the finished queen cells are removed from the colonies in which they are produced and placed individually in mating nuclei. Mating nuclei usually contain small pieces of comb, a limited quantity of honey and pollen and five hundred to a thousand or more bees. The virgin queens that emerge in these nuclei are fed and cared for by the workers in them. It is from these

Capped queen cells about to emerge. These cells have been placed on wooden cell bases. The cell on the right has an excessive amount of wax surrounding it; this may occur during a good honey flow. Note the tips of the cells on the left and right have been removed by the bees to aid the emergence of the virgin queen.

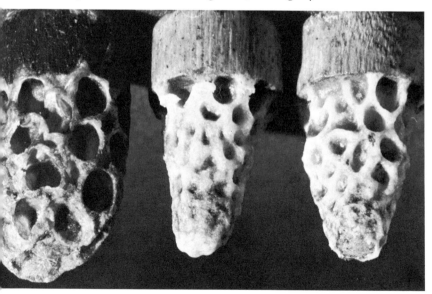

Examining a queen mating nucleus in Florida. This box holds two nuclei, each with about 1,500 bees and a virgin queen; after mating, the queens are moved to larger colonies and the nuclei are used for the same purpose again.

small mating nuclei that the virgin queens take flight and subsequently mate. When the queen rearer notes that the newly mated queens are laying eggs, the queens are placed in queen cages together with five or six worker bees and sold.

Queen rearing requires some special equipment. Because not many people are engaged in queen rearing, the equipment is not made commercially and queen breeders make the necessary special frames, cell bars and queen-mating nucleus boxes. Wax cups, into which young larvae are grafted, are available from the commercial bee supply companies. The same companies also make the queen cages in which queens are mailed; however, the equipment used between grafting and mailing the mated queen is a product of the queen breeder's workshop.

Queen rearing requires considerable attention to detail, especially insofar as removing the ripe queen cells from the colony is concerned, so that the emerging virgin queens do not kill one another. Successful queen rearing requires the development of a market, though the existing bee journals provide sufficient advertising media. The market could be expanded in the north.

Several books have been written on queen rearing. Two of the best, written by Jay Smith and Walter T. Kelley, are out of print; they are available from some libraries and secondhand book dealers. A third book, by Professors Eckert and Laidlaw of the University of California at Davis, is still available. All three of these should be consulted by anyone seriously contemplating rearing queens on a large scale.

Comb-Honey Production

At the turn of the century almost all of the honey produced in the United States was sold as comb honey (honey in the comb). While the favorite size comb-honey section contained thirteen to fourteen ounces of honey, comb honey was also made in two-pound boxes and rarely in quarter- and half-pound sections. Beekeepers produced comb honey because honey sold in liquid form was frequently adulterated with cheaper sugars. The Pure Food and Drug Laws were not passed until 1906 and until then beekeepers had no way of protecting themselves against this adulteration. However, consumers understood that honey sold in the comb was pure.

Following the passage of the Pure Food and Drug Laws, the market for liquid honey increased slowly. During World War I there was a greater demand for honey as the sugar supply in the United States was cut as a result of warfare on the sea. Since it is cheaper and easier to produce liquid honey, many beekeepers began to do so at that time. This effort was intensified for the same reasons during World War II. As a result, almost no one in the United States today makes a full-time living producing comb honey; however, comb-honey production is a favorite pastime for many amateur beekeepers. Where there is a good local retail market, comb-honey production can also be a profitable sideline, but the production of comb honey, generally speaking, costs much more per pound than does the production of liquid honey.

As one examines the beekeeping industry it is difficult to say whether the production of queens or the production of comb honey represents the more difficult aspect of the art. Certainly both queen rearing and comb-honey production require considerable attention to detail. Comb-honey producers must thoroughly understand the biology of the colony and must be in a position to make close examinations of the brood nest during the active honey flow when the sections are being filled.

Successful comb-honey producers reduce the brood nest to one super, one to two days after the start of a major honey flow and at the time when the comb-honey supers are added to the colony. It is known that young queens are much less inclined to swarm than old queens and, since crowding encourages swarming, comb-honey producers try to have young queens in their colonies at the start of the honey flow. Frames containing capped brood are placed in the single brood nest super so that the bees in them will emerge as soon as possible and add to the overall population of the colony. This also provides a place for the queen to lay eggs in an otherwise crowded hive. Even with as many precautions as the beekeeper can take, it is necessary to examine each comb in the brood nest at intervals of seven to eight days during the honey flow. Any queen cells produced by the bees must be removed. While removing queen cells does not guarantee that a swarm will not depart from the hive, it usually prevents swarming in colonies.

During the honey flow the supers must be rotated and the filled sections removed as soon as they are capped. Since the final package is sold to the consumer in the wood that surrounds it in the hive, it is important that the wood be kept clean and that the amount of travel stain on the comb be kept at a minimum.

Comb-honey producers have several problems not encountered by the producers of liquid honey. In years past it was recommended that comb-honey sections be fumigated before they were marketed; today there is no fumigant that is approved for comb-honey fumigation. The only recourse is for the beekeeper to place his comb-honey sections in a freezer for a short while so as to kill any wax moth eggs or larvae that may be present. (See "Wax Moths," page 121.) Premature granulation of honey in the comb can also be a serious problem. It is for this reason that comb honey is usually moved to the market and sold as rapidly as possible. Some honeys granulate more rapidly than others and beekeepers must be aware of the major honey-producing plants in their area. Comb honey is a delicate product and it is not advisable to ship it by routine shipping methods.

Still, there are few things in beekeeping today that can give greater satisfaction than the production of a fine super of comb honey. Not only is there a good demand for comb honey, but comb honey makes an unusual gift. The process of extracting and removing honey from the comb, straining, filtering and pasteurizing it has some effect on the flavor. Since honey in the comb is unprocessed, it retains its delicate flavor and is an unusual, natural product.

Like the processes used to rear queens, those for producing comb honey were developed in the late 1800s. Several books have been written on the subject of comb-honey production, but most of them are unavailable today except through second-

A section of comb honey fully capped. It is difficult to force bees to fill the corners of the combs with honey; thus round sections (Cobana) are becoming increasingly popular for comb-honey production.

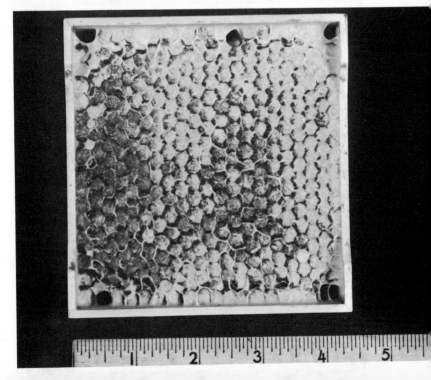

hand book dealers. Several states have issued bulletins on the subject and articles on comb-honey production appear frequently in the bee journals.

Chunk and Cut Comb-Honey Production

Cut comb honey is comb chunks containing two to sixteen ounces of honey. These pieces of comb are cut from much larger combs, usually produced in standard half-depth frames that are about four by seventeen inches. The individual pieces of comb are usually cut with a hot knife and the honey from the cut cells is allowed to drain for up to twenty-four hours before the individual pieces are wrapped in cellophane or plastic. The wrapped pieces of cut comb are normally protected

A section of comb honey only partially capped. Notice that the capping of the honey cells started in this section near the center of the comb. Many bees participate in the capping of each cell of honey.

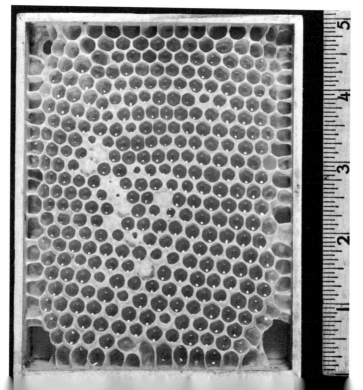

by cardboard for shipping. Cut comb honey has the advantages of comb honey, especially in flavor and appearance, but the production techniques are much easier.

Chunk honey is a piece(s) of cut comb honey in a jar surrounded by liquid honey. Chunk-honey packs have been favorites in the southern states for decades. However, most of the honeys produced in the southern states are much slower to granulate than those produced in the northern states. Thus, a chunk-honey pack would have a shelf life of four to six months in the south while in the north a similar pack might have a shelf life of only a few weeks. A granulated chunk-honey pack has little appeal to the average purchaser.

Producers of combs of honey to be cut into smaller pieces usually follow the same management techniques used by comb-honey producers; however, it is not necessary to pay quite so much attention to detail. The fact that there are no dividers or fences between the combs, as there are in comb-honey supers, makes the chunk-honey supers less crowded and swarming is not so much of a problem. Chunk-honey producers do not need to crowd their colonies to force the bees to fill every section. Pieces of comb not perfectly filled may be trimmed or not used at all without too much monetary loss. In the production of comb honey a producer cannot afford to discard too many sections just because of the extra cost of the wood and labor involved in preparing the supers.

It is possible to produce a few frames of chunk honey by placing the frames in the center of an ordinary extracting honey super. If this is done, the super should have only light-colored combs on either side of the frames in which chunk honey is to be produced; otherwise, the chunk-honey frames will be travel-stained and will not have a good appearance. Whenever a light-colored comb is placed next to an old, dark comb, it can be observed that bees just walking from one comb to another carry enough residue on their feet to stain the new comb. Removing the newly filled comb, in the case of the production of both comb honey and chunk honey, is advisable for the same reason.

Showing Honey

Honey shows are becoming increasingly popular everywhere in the country. Not only is a honey show at a fair an excellent method of advertising honey, but often sales from the show booths can be profitable. Also, honey shows offer an opportunity for beekeepers to improve their methods of packing and preparing honey for market through observation of what others are doing.

Unfortunately, both the rules and the judges at various honey shows vary and this makes it difficult for those preparing entries to know how to proceed. Most judges agree that foam should not be present on honey in shows and they also agree on the quantity of foreign material and wax particles that should be allowed. Some shows indicate in their rules that the honey should have a predominate flavor, and often honeys from mixed sources do not qualify under these circumstances. Judges do not agree as to the number of points to be assigned to the physical appearance of a jar or container in a honey show or the importance of this item in judging. More important, some judges pay great attention to the quantity of moisture in the honey while other judges do not. All judges will disqualify honey that contains more than 18.6 percent moisture, but beyond this there is little agreement. Since the beekeeper has less control over the moisture content of the honey than practically any other of its characteristics and qualities, some judges feel that moisture content should not be so important a consideration.

The attached grading chart shows the distribution of points by one group of qualified judges. Where possible, it is advisable to force the judge to use a grading card and to assign points in each category. It is only by doing so that the beekeeper can learn where errors were made and where emphasis should be placed to improve the quality of future entries. It is also helpful for both the participants and the judge if the rules for the show are printed and available to everyone well before the show. Some of the items that should be predetermined are the size and type of jar and cap, the number of jars for each entry (three is the usual number), when the entries are due, and how moisture content is to be judged.

SCALE OF POINTS FOR JUDGING HONEY

Liquid Honey *Points*

1. Appearance, suitability and uniformity of containers 10
2. Uniform and accurate volume of honey 5
3. Absence of granulation, impurities and uniformity of the honey entry 25
4. Color and brightness 20
5. Flavor and aroma 20
6. Density 20

100

Granulated Honey *Points*

1. Appearance, suitability and uniformity of containers 10
2. Uniform and accurate volume of honey 5
3. Firmness and uniformity of set, absence of froth and impurities and general condition of honey 30
4. Color 15
5. Flavor and aroma 20
6. Texture of granulation (smooth and fine) 20

100

Comb Honey in a Standard Section *Points*

1. Suitability and uniformity of sections 15
2. Uniformity and completeness of filling 25
3. Cleanliness and general appearance of section 20
4. Cappings and uniformity of whiteness 20
5. Quality of honey 20

100

Honey Display	Points
1. Educational and/or advertising value	40
2. Attractiveness and arrangement	30
3. Appearance and quality of honey	30
	100

Beeswax	Points
1. Color (straw or canary yellow)	30
2. Cleanliness (free from honey and impurities)	35
3. Uniformity of appearance	20
4. Freedom from cracking and shrinkage	15
	100

Migratory Beekeeping

Men have been practicing migratory beekeeping for several thousand years. The Egyptians apparently moved their bees on barges on the Nile River when their civilization was at its peak. It has been shown that colonies of honeybees can be moved long or short distances without harm to the bees, brood, or stored food. It is necessary to move colonies more than two or three miles; otherwise the field bees will fly back (drift) to their original location.

The rapidity with which field bees adapt to new foraging conditions is quite remarkable. While many of the bees in a colony tend to be quite angry and very prone to sting for several hours immediately after the colony is moved to a new location, many of the worker bees will start to forage immediately. Observations by several people indicate that foraging bees can return to newly moved hives with pollen loads within fifteen minutes after being moved to a new location.

The danger in moving a colony to a new location is not chilling the brood, but overheating the colony. When colonies are being moved from one location to another a cluster will form in each hive if there is danger of chilling the brood.

Within the cluster heat is generated and in this way the brood is protected. However, bees cannot guard against overheating as easily. If the bees are unable to move large volumes of air through the colony it may overheat, the excess heat being produced by the bees themselves in their unsuccessful attempt to cool the hive. In hives which overheat the combs may melt and the bees and brood die. For this reason it is safest to move colonies during cool weather and/or the evening hours.

The safest way to move a colony of bees is to smoke it gently and to place it with an open entrance on a truck and move it to the new location. It is an interesting fact that the vibration of a truck bed will usually calm the bees and inhibit most flight. Thus, there is not too much danger of too many bees being lost from the colony while it is being moved. A major problem with open-entrance moving in warm weather is that large numbers of bees may come out and cluster on the outside of the hive, making it difficult to remove the hive from the truck bed. Likewise, in open-entrance moving some bees may escape from the truck if it is forced to stop at a red light or stop sign, and it is possible that people in the vicinity may be stung. Beekeepers who move their colonies open-entrance usually cover the entire truck with a large plastic or wire screen to prevent the bees from escaping from the truck.

While open-entrance moving is favored by many beekeepers, the more conventional method is to use a top screen and some type of entrance screen. The top screen is usually made in the shape of a super and has a rim about two inches deep. This rim is covered with some type of wire screening; eight-mesh hardware cloth is toughest and works best for this purpose. If the colonies are placed one on top of another, it is necessary to place slats, at least two inches square, between the piles of colonies. Several types of entrance screens may be used including porch screens and tuck-in screens. The so-called tuck-in screens are much easier to use and require less time. However, the porch screens are more effective in keeping the bees quiet and presumably aid in giving the colonies better ventilation. Even where large screens are used, some beekeepers wet their colonies externally, and even internally, with water from a garden hose during the heat of the day.

It is often possible to move one or two colonies of bees without stapling or otherwise nailing the supers and bottom board and cover together. If bees have used large quantities of propolis and if the parts of the hive have not been broken apart in recent weeks, often this gluelike material is sufficient to keep the parts together. However, most beekeepers who practice migratory beekeeping use either steel strap or hive staples to tie the parts of the hive together. Hive staples are usually larger than ordinary staples, being about two inches long, and they do less damage to the wood. It is possible to fasten the parts of the hive together with wooden slats, and this is satisfactory if only a few colonies are being moved; however, wooden slats usually take up too much room when large numbers of colonies are being moved from one location to another. Beekeepers who move many colonies over long distances usually build special bottom boards for their colonies that are not so cumbersome and bulky as the standard bottom board. Usually these bottom boards take the form of a combination bottom-board cover and may be easily nailed in place while the colonies are being moved. If large numbers of colonies are moved long distances, it is advisable to limit the size and weight of the units as much as is possible and practical.

When several colonies are moved at one time, the entrances of colonies should be opened at dark, but this may not always be practical. If the entrances of a number of colonies are opened at the same time and on a warm day, when bee flight would be encouraged, many bees may fly from their colonies and be lost and drift to other colonies. It is usually, though not always, the colonies on the end of a row that "pick up" the greatest number of bees under these circumstances. Placing colonies as far apart as is practical and painting the hive bodies different colors are two techniques used to reduce drifting under these and other circumstances.

Removing Unwanted Bees from Buildings

Honeybees commonly nest in the side of a house, barn or shed. The natural home of a honeybee colony is a hollow tree or cave; therefore, any dry cavity, such as a space between the studding of a house, is a desirable nest site. Colonies have lived in houses for ten to twenty years without bothering the inhabitants or otherwise being a nuisance. So long as the nest remains active and the bees fly over the heads of people in the vicinity, there is no problem; however, should the bees die, the comb, honey and pollen become attractive to a great variety of animals, such as ants, cockroaches and mice. Since this is true, it is best to remove the bees, and the nest, when it is convenient to do so.

Honey is hygroscopic (moisture-absorbing) and this could cause trouble if the bees in a colony in a house should die. Honey will absorb moisture from the air, if not protected by bees. If this takes place, the honey can ferment, break the comb and run down on the ceilings and walls. Such honey-soaked areas are often difficult to clean. In hot weather, combs unprotected by bees can melt and release the honey.

It is possible to save a colony of bees that is nesting in a building, but it is not recommended unless the person doing it has a great deal of experience in handling bees. It is far better to kill the bees with an insecticide. In either case, it is necessary to remove all the combs and honey, wash the area with soap and water, and fill it with insulation or some other material, so that bees cannot reoccupy the old nest site. Bees like to nest where bees have been before because of the odor left behind. Old nesting areas may remain attractive, because of the residual odor, for several decades.

When a colony is killed with an insecticide, the comb, honey and dead bees should be burned or buried; most insecticides are somewhat toxic to man. The honey should not be eaten. Almost any insecticide applied to the nest in excess will kill the bees; contrary to popular opinion, DDT is not very toxic to bees.

Where the colony is between the studding in the walls of a building, one of the simplest procedures is to drill three or four holes in the wall above the nest and introduce about one-quarter or one-half of a cup of an insecticide solution into each of the holes with a small funnel and rubber tubing. All of the bees will be dead in a few hours. However, it is best to wait about twenty-four hours before removing the dead bees, honey and comb. If the nest is treated in this manner, the honey and comb should be burned or buried; under no circumstances should it be used as a food for humans or bees.

Transferring Live Bees from an Inaccessible Nest

The best time to trap bees out of a building and remove the honey is at the beginning of the main honey flow, which occurs in the northeast about the middle of June. They can be trapped out at the beginning of the dandelion honey flow in the spring or the buckwheat honey flow in late summer, but the beginning of the clover-honey flow is usually the best time. The following procedure is recommended:

1. Close all the flight holes in the building except one.

2. Build a strong temporary platform near this hole, large enough to hold a standard bee hive.

3. Fasten a bee escape over the single flight hole that has been left so that the bees may leave the nest but not return to

it. Factory-made bee escapes can be purchased from any of the bee supply houses or dealers.

An efficient type of homemade bee escape can be made by forming a piece of ordinary cloth screen into a cone. A cone of twelve or more inches in length should be used, but the opening at the small end should be just large enough to let one bee at a time pass through. The correct diameter of the small opening is about three-eighths of an inch. The diameter of the base or the large end of the cone, which is fastened over the opening of the nest, should be not less than six inches in diameter but it may be larger without lessening the efficiency of the escape.

4. A small colony, consisting of two or three frames of sealed brood and bees with a caged queen in a standard hive body, with a bottom board and lid, should be placed on the platform. The platform should be constructed not more than ten feet below or away from the nest. It is advisable to fasten the hive to the platform. The hive body should be given its full complement of drawn comb, and one or more supers of comb should be added, as needed, to provide sufficient room for the queen and bees.

5. The queen should be released from her cage about three days after the bee escape has been put in place. When the bees find that they are unable to return through the bee escape to their old nest, they will enter the new hive prepared for them. In about a month, practically all the bees will be in the new hive on the platform and the queen in the old nest will ultimately die from lack of sufficient bees to care for her.

6. After a month, very few bees will remain in the old nest and it should be possible to remove the siding or boards protecting the nest and remove the combs.

Salvaging Honey from Old Combs

People are often disappointed when they taste the honey in

old black comb from a nest in the side of a building or from a bee tree. Honeycomb becomes black after a few years because of pollen stored in the comb and because of propolis, the beekeeper's term for gums and resins bees collect and add to their combs to strengthen them. Both pollen and propolis usually have a bitter taste. Old comb containing honey should be broken and crushed and the honey allowed to drain out of it. This may take a day or more. The comb may be squeezed slightly but if pressed too hard, some of the bitter components may be added to the honey. After the honey has been removed, the old comb should be burned or buried. The honey may be strained through cheesecloth before it is put on the table.

Honeycomb that is white or light yellow in color is new comb and may be eaten along with the honey. Frequently there will be white or yellow comb along the sides and on top of the nest. Beeswax is digestible and may be eaten or chewed like gum. It is possible to recover the beeswax in combs removed from a building if they are not contaminated with insecticide.

Removing Nests in Cold Weather

If it is not convenient to trap out the bees, and if access to the colony is possible without causing serious damage to buildings or trees, the bees and honey can be removed during freezing weather when there is no brood in the combs. Honeybees soon become inactive and die when exposed to cold air. The bees can be brushed from the comb and the portions containing honey can be saved for table use. The removal of bees from trees, boxes or from other unsuitable containers can better be done during warm weather in order to salvage the bees as well as the honey.

XI

Major and Minor Honey Plants

Honeybees harvest what would otherwise be a wasted sugar supply. Beekeepers do not grow honey plants to produce nectar; they depend upon natural sources of nectar and pollen both to sustain their bees and to produce a surplus that man may harvest. Our present knowledge of nectar-producing plants indicates it would not be profitable to grow plants for nectar production alone. If one can grow a crop that has cash value and at the same time produces nectar, such a venture might be profitable.

Over half of the major sources of honey in the United States are plants not native to this continent; many of these are weed plants and were brought here accidentally. In the north most of the surplus nectar is produced by ground flora, that is, plants that grow to a height of one to six feet. Generally speaking, in the north lighter honey is produced. In the tropics and many subtropical areas, most of the nectar is produced from trees; in general, tree honey is darker than that produced by ground flora. However, these are general rules and there are many exceptions.

In some states the honey flora have been carefully examined and articles have been published outlining the major and minor honey-producing areas within the state. Information about articles and bulletins published on this subject can be obtained by writing the state apiculturist at the state college. Often moving an apiary a distance of two to four miles can make a difference between a large and a small crop of honey; for this reason it is important that the beekeeper be familiar with the plants in his area. While management of colonies is stressed for those interested in honey production, even that cannot make up for a lack of natural sources of pollen and nectar.

The Scale Hive

The best way to determine when a honey flow starts and ends is with a colony of bees on scales. A scale hive also provides information about spring buildup and the condition of the colony prior to winter. Scale hives are often misleading because they usually receive the best care of any colony in the apiary. If anything goes wrong with a scale colony, it is immediately evident to the beekeeper and it usually receives special attention.

Scale-hive records kept over a period of five or ten years are especially valuable in pinpointing seasonal changes. In the more northern areas it is possible to predict within a few days the date a honey flow will start; in the south this is not always so easy. In Florida, for example, oranges may start to produce nectar as early as late February or as late as the first few days of April. Long-range scale-hive record keeping is the only way a beekeeper can predict what will happen. With a thorough knowledge of the major and minor honey flows in an area it is much easier to prepare a practical management schedule.

The Major Honey Plants

The major honey plants are those which produce nectar in sufficient quantity that the bees are able to store a surplus above their needs. This surplus may be harvested by man. Whereas in any one particular area bees may collect nectar from a hundred or more plants, there are usually only two to four that the beekeeper considers major honey plants. In wholesaling honey, beekeepers speak of and sell their honey in terms of these major honey-producing plants. Occasionally honeys are sold in the retail market under the name of the plant from which the nectar is produced; however, on the retail market most honeys are blended.

Not all of the major honey-producing plants yield nectar equally well in all areas. While there have been very few thorough studies on the subject of nectar secretion, there are indications that alfalfa, for example, produces the greatest quantities of nectar on well-limed soils. Some beekeepers feel that buckwheat does not produce nectar in quantity on limed soils but that it produces best on hilltop farms and on the poorer soils. In addition to soil type, nectar production is highly dependent upon weather conditions.

As agriculture changes, so does the availability and quantity of nectar-producing plants. The change in agricultural practices in central and western New York State is an excellent example. From the time that central and western New York State were first settled until the early 1950s, buckwheat was one of the most common grains grown. Buckwheat seed was used for food for animals and to make flour for pancakes. However, shortly after World War II buckwheat seed production started to decline. Buckwheat grain is high in fiber and low in protein; it does not compete well as a feed grain with wheat and corn. Thus, buckwheat honey became a scarce commodity because so little buckwheat grain was grown. About the same time it became evident that more farms were going out of business in the same area. As fields were abandoned they grew up to goldenrod. Goldenrod has always been known as a good source of nectar, but it was usually not available in sufficient quantity for beekeepers to secure surpluses of goldenrod honey. Shortly after World War II goldenrod became a more important honey plant. At the same time the agricultural colleges in the north were seeking better forage plants for dairy cattle. It was decided that alfalfa was the best of the sundry forage plants available and it was recommended that it be grown on New York State farms and that heavy applications of lime be made where the soils were not sufficiently sweet to support the growth of alfalfa. Starting in 1957 it became evident that alfalfa was a major honey-producing plant in many parts of New York State. Prior to this time alfalfa had been an important nectar producer in some local areas. In fact, as early as 1927 Dr. Phillips, then at Cornell University, reported that alfalfa was a major honey-producing plant in a small area around Syracuse. Occasionally, when it is wet during August, alfalfa does not produce a surplus of nectar; but generally speaking, alfalfa is the major honey-producing plant in New York State and re-sponsible, in the average year, for about half the honey pro-duced in the state. At the same time, the number of acres of goldenrod appears to be increasing, though in some areas the goldenrod is decreasing as the weed-infested fields grow up to small trees and brush. The increased growth of forests in the area will undoubtedly result in more basswood honey being produced.

Beekeepers who have a few colonies in their backyard may not be able to take advantage of special and changing agri-cultural conditions in their area. However, beekeepers who are planning to establish outyards are advised to inspect soil maps and to consult the county agricultural agents in their area to determine the agricultural patterns. As agriculture changes, so will the available nectar. Beekeepers in large city areas may prefer to establish their colonies near parks or to search out villages and towns in the vicinity that have made extensive plantings of special trees such as basswood or locust. In some areas towns have lined streets with such trees and collectively these can produce large quantities of honey.

The following are some of the major honey plants in the United States and the general areas in which they grow:

ALFALFA (*Medicago sativa*)

Alfalfa has been an important honey plant in the western states for many years, especially on irrigated land. It has had limited use as a bee forage plant in the east until recently.

In the 1950's alfalfa became the primary source of honey in New York State and in parts of certain other northeastern states. During this time alfalfa became the major forage plant on dairy farms in this part of the country. Many commercial beekeepers now seek out large dairy farms as locations for apiaries. Alfalfa yields best on well-limed and well-drained soils; these are usually the better soils in the northeast. Little nectar is produced by alfalfa flowers in June or early July on the first cutting; it is the second and third cuttings, in late July, August and early September, when the honey crop is secured.

Alfalfa honey is light in color and mild in flavor. It is typical of the high-quality honeys produced by the legumes. Alfalfa honey has a delicate flavor and is packed both as a pure product and as a blend with other, slightly stronger-flavored honeys. The rapidity with which nectar is secreted by alfalfa makes it a very desirable plant for beekeepers interested in comb-honey production. Alfalfa honey does not granulate too rapidly in the comb and the cappings on the combs are white and give a good appearance.

Alfalfa used for hay or silage should be cut when the flower buds are still tight and unopened. However, most farmers are behind in their work and much alfalfa is cut at a later stage. In fact, if alfalfa were cut at the recommended age, New York State would produce no alfalfa honey. At the present time it is one of New York's major sources of nectar.

ASTER (*Aster* sp.)

Aster, like goldenrod, is widely distributed and native in North America. Most authorities on honey plants indicate that aster honey is gathered at the same time as goldenrod, which is true only in part. In fact, aster honey may be responsible for the tendency of goldenrod honey to granulate so rapidly since it is known that pure aster honey does granulate very rapidly.

In most of the northeast goldenrod is killed by a frost in mid- or late September. Aster is not affected by the lighter frosts and it takes unusually cold weather to kill it. Rarely, after a strong frost that has killed goldenrod, there will be an excellent honey flow from aster in the northeast. This probably does not occur more than one year in ten. In the Ithaca area there have been aster flows that yielded as much as 60 pounds of surplus honey after the goldenrod honey had been removed from the colonies. In the cool fall months aster honey tends to granulate very rapidly and can granulate as soon as a week or ten days after it is placed in the comb by the bees; for this reason it must be extracted as soon as the honey flow is finished.

Aster honey is white, and has a delicate, perhaps minty flavor that is unusual. Because aster honey is usually mixed with goldenrod honey, it is unknown to most people except a very few beekeepers who have made a special effort to produce it.

BASSWOOD (*Tilia* SP.)

Basswood is a common native American tree that yields large quantities of a white aromatic nectar. Several species of basswood are found in this country including some imported from Europe. The trees flower during the first or second week in July and nectar is produced for one to two weeks. The honey flow from basswood is usually intense. Basswood is an erratic nectar producer, not producing every year and subject to weather conditions. A heavy rain or a few days of inclement weather usually stops nectar secretion.

Basswood honey is light in color and has a slightly minty flavor. It makes excellent comb honey; there is very little pollen available at the time and the cappings on the sections are usually white and clean. Basswood was known as a fine nectar producer by the early settlers in this country, but as the land was cleared for agricultural use, the number of basswood trees declined (it is also a desirable lumber tree). Recently, as poor agricultural land has been abandoned in certain parts of the country, basswood has become more common and in many areas of the northeastern United States large crops of basswood honey are produced.

Both American and European species of basswood have been used by certain towns and villages as ornamental trees. In certain towns in New England, where large numbers of these trees line the roads and highways, good crops of basswood honey are obtained. Many people not familiar with basswood honey feel that the flavor is too strong for common acceptance; because of its light color, basswood honey is used largely for blending in table packs.

BUCKWHEAT (*Fagopyrum esculentum*)

A few decades ago buckwheat was the major honey-producing plant in the northeastern United States. Buckwheat was brought to this country from Europe, though it is Asian in origin. It was a popular plant for farmers in the northeastern states because it could be sown late in the year on the wet and poorer soils and still produce a crop of seed before frost. Because buckwheat is a poor grain for feeding livestock, it has gone out of favor during the past several years; it is high in fiber and low in protein. Small quantities of buckwheat are still grown in parts of western New York and Pennsylvania and other isolated areas. Buckwheat honey, which was once surplus on the market, now commands a high price, not infrequently a price above that of the premium honeys, because of its scarcity.

Pure buckwheat honey is coal black and has a very strong flavor. Experimentally in our laboratory we have mixed one part of buckwheat honey with seven parts of fall flower honey and in the resulting mixture the buckwheat flavor still dominated. Much of the honey on the market today labeled buckwheat honey actually contains only 15–25 percent buckwheat, so strong is its flavor.

Most of the buckwheat honey produced is gathered in the month of August. Buckwheat may be sown any time up to the first week of July and produces nectar usually in about six weeks after the seed is sown. It is difficult to discuss the merits of strong-flavored honeys; among those who eat them regularly they are delicacies. People who taste buckwheat for the first time are often repelled by it.

THE CLOVERS (*Melilotus alba, Melilotus officinalis, Trifolium* sp.)

Collectively the clovers have been by far the most important American honey plants. This is especially true in the north central states. In the northeastern United States the clovers are important only in limited areas, usually those with good soil and a high natural lime content. In all of the northeast clovers are important minor honey plants.

Most of the clovers start to flower sometime in mid- or late June and continue to yield nectar through about the first of August. Clover honey is, generally speaking, light in color and mild in flavor. Thirty to fifty years ago clovers were probably far more common in the northeast than they are today; in fact, this is true of much of the northern United States. As a result, many of the early bulletins and texts on beekeeping refer to the clover honey flow as being the major one in many areas. As agriculture has changed, this has become less true.

GOLDENROD (*Solidago* sp.)

Goldenrod is one of the more widely distributed honey plants in the United States. There are many species and varieties, most of them native to this country. Because goldenrod is so widely distributed, it may be considered a minor honey plant in almost every beekeeping region of the United States; however, in many areas in the northeast, especially where farmlands are being rapidly abandoned and goldenrod is growing profusely, it is a major honey plant.

In the Ithaca (New York) area, for example, we can depend on a goldenrod honey flow about two years out of three. The yield will vary from fifty to one hundred pounds under normal circumstances. Goldenrod nectar has been collected as early as the first of August and a major honey flow has started as late as September 10 in this area. It is much more difficult to predict when a goldenrod honey flow will take place than in the case of most other honey flows.

Most goldenrod honey is sold to the bakery trade. Goldenrod honey has a deep yellow, even golden color. It normally has a low moisture content. Goldenrod honey granulates very rapidly, usually within two to three weeks after it is extracted. Goldenrod honey is widely used in New York State as a wintering food for bees. It appears to be satisfactory in this regard though colonies that have been examined in December and January have shown large quantities of granulated honey and it is undoubtedly difficult for the bees to remove and to use this efficiently during the cold winter weather.

LOCUST (*Robinia pseudoacacia*)

The black locust, sometimes called the honey locust, is native to the northeastern United States. It was carried throughout the country many decades ago. Black locust logs are slow to rot and decay and have been in great demand for fence posts in past years. In many parts of the country one will find small plantations of the trees which were planted for the express purpose of producing fence posts.

Black locust usually flowers in early June and yields a copious supply of nectar. The honey is light in color and mild in flavor, not unlike that of clover. Black locust and basswood are probably trees that beekeepers could encourage to be planted in parks, along roadsides and elsewhere.

ORANGE (*Citrus* SP.)

Under favorable circumstances citrus probably produces more nectar per acre than any other major honey plant. It is often possible to shake a large drop of nectar from a single orange flower. Under such circumstances a bee can gather a load of nectar from one or only a few flowers. All citrus honey, whether from tangerines, oranges, tangelos or grapefruit, is sold under the name of "orange honey."

Interestingly, not all citrus requires cross-pollination to set fruit. Thus, it is difficult to say why so much nectar is produced. Generally speaking, plants and animals do not waste natural energy, and why orange produces so much nectar is one of nature's paradoxes. Only a few years ago it was discovered that some of the new varieties of tangelo grown in Florida require cross-pollination. As a result, citrus-fruit growers are interested in having bees in their groves; this was not the case a few years ago.

Citrus does not produce nectar in all the locations where it is grown. Sometimes, too, areas vary and one cannot always depend on even the good nectar-producing areas to produce nectar every year. Citrus honey is variable in color and flavor. The honey may be light in color (white or water white), though more commonly it is amber or light amber. Some citrus honey has a very strong flavor and it is often blended with milder honey.

The time when citrus flowers will yield nectar is also variable. In Florida, for example, the honey flow can start as early as late February or as late as early April. Citrus trees may start to flower, with perhaps 5 to 15 percent of the flowers

Removing supers of honey in a typical commercial apiary in an orange grove in Florida. The brood nests in these colonies occupy the lower two hive bodies; the honey is stored in the supers above and it is these the beekeeper is removing and placing on his truck. This picture was taken in early April.

open, and then remain in this condition producing no nectar for several weeks. Moisture and weather clearly dictate when, how long and how good an orange honey flow will be.

RASPBERRY (*Rubus* sp.)

Raspberry is a major nectar producer in limited areas in the northern states, but a very important minor honey plant in most of this same area. Raspberries grow well on poor soil, though they have more luxuriant growth on good soil. They are common in mountainous areas and appear to be especially abundant around old foundations and on sites where farm buildings once stood. In the limited areas where raspberries are cultivated, they are also important yielders of nectar.

Raspberry honey is light in color and mild in flavor; many

people consider it one of the few honeys superior to light clover honey. Raspberry nectar secretion appears to be less affected by adverse weather conditions and the plants abound with bees whenever and wherever they are in flower. Raspberries—and the related blackcaps and blackberries, which are also probably good nectar producers—start to flower in early June, but it is usually mid-June before any surplus honey is produced. Some beekeepers tell of migrating to the extensive raspberry acreages that exist in the mountainous areas of the northeast, especially the Adirondacks. However, moving bees for this nectar source has been less popular in recent years. Yields of fifty or more pounds per colony have been recorded. Because of its unusual flavor and taste quality, there has been a good, though limited market for raspberry honey, which is sold under its own name.

TULIP TREE (*Liriodendron tulipifera*)

The tulip tree, which often grows very large, is a major honey plant in Pennsylvania, the most southern parts of New England and south to the Gulf states. The trees bloom in April and May and usually produce nectar in great quantity. An occasional tree is found in New York State, but it is not a major honey plant there or in central and northern New England.

Since the trees flower early in the year, bees are often not in a sufficiently strong condition to harvest a surplus. The honey is a light amber color and is a good quality for the table. Yields of one hundred and more pounds have been recorded.

WILD THYME (*Thymus vulgaris*)

Wild thyme is a major nectar-producing plant in parts of the northern and central Catskill Mountains, Albany County of New York State and to a lesser extent in the Berkshire Mountains in Massachusetts. Wild thyme is a plant from Greece, probably imported in the wool of sheep. In Greece, as in the United States, wild thyme covers the hillsides, causing them to be a reddish purple in late July, August and sometimes in September. The so-called purple hills of Greece are so named because of this plant. Likewise, in Greece wild thyme is known as an important source of honey. Undoubtedly, at the height of the Greek and Roman civilizations, when honey was in great demand, it was wild-thyme honey that was "the nectar of the Gods."

The wild-thyme plant is shallow-rooted and grows well on the poorer soils in the northeast. It has a strong odor and, when in flower, the odor from a large acreage may be detected for hundreds of yards and even greater distances. Because the plant is shallow-rooted, it is dependent upon weekly rains for continued nectar secretion. The plant begins to flower in mid-July, and nectar can be produced until frost if there is sufficient water.

Wild-thyme honey has a reddish color and a strong flavor. It is highly prized in the area where it is produced and it is thoroughly enjoyed by those people who prefer strong-flavored honey. Beekeepers not familiar with it say that it should be sold to the bakery trade only, as it is too strong for the average American palate. Comb-honey producers find the wild-thyme plant an excellent one. Wild thyme produces little pollen and as a result the sections of honey have a fine appearance. Some beekeepers in New York State use the summer-flowering wild thyme as a buildup plant for nuclei that are carried into the area for that specific purpose.

The Minor Honey Plants

In any given area in the northeastern United States there are probably two hundred or more flowering plants from which bees collect pollen or nectar or both during the year. These plants are important for the day-to-day success of a honeybee colony, and without them beekeeping would not be possible or profitable. It is difficult to say precisely what a minor honey plant is. Given the proper weather conditions and a sufficient acreage of a given plant, it is entirely possible that what is considered a minor plant can be a major one.

In the early spring there are a series of so-called minor honey plants including willow, maple, dandelion, yellow rocket and apple, all of which produce nectar and pollen in fair quantity and are important for building the population of a bee colony in the spring. Rarely, beekeepers have reported securing a crop of honey from all of these plants. Probably, if the colonies were of sufficient strength and the weather satisfactory, it would be entirely possible to produce a surplus from any one of these plants. Since most of the plants that flower in the spring also yield large quantities of pollen, the resulting honey is often yellowish in color and cloudy in appearance. This is especially true of dandelion and yellow-rocket honey. As a result these honeys are not especially desirable for the retail market.

In many parts of the United States it has been found that keeping bees in or near cities or large villages can be profitable insofar as the buildup of colonies is concerned. While it is not recommended that beekeepers try to grow plants for the production of pollen and nectar, it has been observed that collectively the flowering shrubs, etc., in an urban area can be important. This also means that beekeepers might encourage the use of certain trees or shrubs where they have an opportunity to make recommendations. This would be especially important in the beautification of parks and roadsides. It can also be helpful in areas where wasteland is being reclaimed.

Writing about minor plants is difficult, if not impossible. However, this is an area of study that can be followed profitably by local bee clubs. Often such clubs have prepared bulletins enumerating the plants of value in their area as well as the trees and shrubs their members might encourage to be planted.

Honeydew Honey

Honeydew honey is an important honey in Europe, where it commands a high price. In North America honeydew honey is usually disliked and blended with other dark, strong honeys that are sold to the bakery trade. Honeydew honey is of two types. Most of it is an aphid secretion collected by bees. Aphids, better known as plant lice, which feed on certain trees, secrete a sweet, sticky substance high in gums and dextrins. The Black Forest in Germany is one well-known honeydew-producing area.

Another type of honeydew honey is produced by extrafloral (nonfloral) nectaries. Extrafloral nectaries occur on plant stems and leaves and produce a nectar that resembles normal plant nectar. However, most plant honeydews are dark in color and strong in flavor. Cotton is a plant that often yields extrafloral nectar in quantity. The reason some plants have nectaries other than in flowers is not clear. It has been suggested that ants may be attracted to the plant by the nectar; subsequently, any animal, such as a horse or cow, which might eat the plant, would be stung by the ant. In such a case the ants substitute for spines on the plant and offer it some protection. Plant and animal protection systems take a great variety of form.

Privet honey—honey produced from the white flowers on common privet hedges—is often confused with honeydew honey. Privet honey is dark, even cloudy and has an almost foul taste. Like honeydew honey, it is not too common. Most beekeepers in North America think this is fortunate; however, some people enjoy the darker, stronger honeys.

XII

Pollination

Pollination is the process whereby pollen—the male germ cells —is moved from the male parts of a flower to the female parts. Self-pollination occurs when pollen is transferred from the male to the female parts of the same flowers, flowers on the same plant, or flowers on plants arising from the same scion; cross-pollination occurs when pollen is transferred to the female parts of a flower on another plant of the same species but not the same scion stock. Since plants are stationary, some outside force must act on their behalf if cross-pollination is to occur.

Cross-pollination is much more desirable than self-pollination. The resulting offspring are much more vigorous when cross-pollination occurs. Self-pollination is discouraged in nature by several means: the male and female parts on the same flower may be sufficiently far apart that the pollen cannot easily come in contact with the female parts of the flower without an outside agent; the length of the male parts is much shorter (or longer) than that of the female parts so that again pollen must be carried from one to the other; the male and the female parts of a flower may mature at different times, even in the same flower, thus necessitating the movement of pollen from one flower to another; some plants have separate male and female flowers; a few plants carry the separation of the male and female parts even further—there may be male and female plants.

A few pollens are light in weight and can be carried by the wind. Some pollens are large and can float and be carried on water surfaces; in a few cases rain may cause pollen to splash from the male to the female parts of a flower. However, many plants are pollinated by insects and the pollen is carried from one part of the same flower to another part, or from one flower to another flower on the same or another plant. Certain insects may redistribute pollen already carried to a flower other than that on which it originated.

Plants produce nectar for the purpose of attracting insects to them so that the flowers can be pollinated. In this regard nectar can properly be called a "bribe." Many insects are attracted to flowers and one need only to observe the activities of insects on flowers to understand this. But only bees have specialized to the extent that they are wholly dependent upon flowers for all of their food. Bees, considered as a group, obtain their carbohydrate from nectar and their fat and protein from the pollen. It is for this reason that some plants produce great quantities of nectar and pollen. It has even been suggested that there may be some co-evolution of flowering plants. It appears that plants that flower in the spring produce more pollen than nectar; at this time of year protein is needed in large quantity to rear young bees. In the fall it appears that many plants produce more nectar than pollen; in the winter bees need large quantities of honey to sustain themselves.

Solitary and Semisocial Bees

There are over twenty thousand species of bees, including the honeybee, in the world. Some of these are solitary insects and the sexes come together only for the purpose of mating; often the female does not live long enough to see her offspring. There are also semisocial and subsocial bees, which may produce colonies of fifty to several hundred individuals; the bumble bee is an example. In the case of these semisocial bees, the females usually mate in the fall and overwinter as individuals, starting their nests anew each spring. Of all the pollinating insects known to man, the most widespread and easily managed is the honeybee. While two species of solitary bees are used for alfalfa pollination under specific circumstances in certain western states, only the honeybee may be moved rapidly and easily by man into an intensified agricultural situation. For this reason, in certain parts of the world and in the case of certain crops, agriculture has become dependent on honeybees. In addition to their value as agricultural pollinators, honeybees are important for the pollination of the many fruit, seed and nut crops used by wild life. Honeybees are also responsible for the pollination of many wild flowers.

While there are many insects that carry pollen from one flower to another, there are reasons this is not done and why it is necessary to use honeybees for pollination. A shortage of wild bees may be credited to one of the following reasons: an intensified agricultural system where large acreages exist without necessary nesting sites or the alternate sources of food when the crop being pollinated is not in flower; adverse weather, especially in the early spring when the flowering time may be short and bees are needed in large numbers; an elimination of nesting sites either through the plowing and cultivation of sandy, ground-nesting areas, or the elimination of hedgerows and the hollow twigs and branches for bees that build their nests above ground. Contrary to popular opinion, there is no proof to show that modern-day pesticides have had an adverse effect on the populations of wild pollinators in the United States. Examination of flowers in and around the apple

orchards in New York State shows that there are large populations of wild bees even in these areas. However, pesticides continue to be a popular explanation for the shortage of pollinating
insects in certain areas of the country.

In most parts of the United States solitary and subsocial bees
outnumber honeybees about two to one. A few of these wild
bees, such as the bumble bees, of which there are several
species, and the carpenter bee, are large. However, most wild
bees are small; on a weight basis the average solitary bee
weighs less than 10 percent of the weight of a honeybee. In
the case of some flowers, these small bees are very efficient
pollinators; but a plant such as bird'sfoot trefoil or alfalfa requires a bee large enough to trip the flower in order that cross-
pollination be accomplished. A large bee such as the honeybee
has the added advantage of being able to carry a greater quantity of pollen on its body. In the case of certain flowers, this
added pollen can be of consequence in pollination.

The Proper Size Colony for Pollination

It is generally agreed that for the pollination of most agricultural crops in the United States colonies of bees in two
supers (boxes or hive bodies) should be used. This means the
use of two standard ten-frame Langstroth hive bodies. Hive
bodies that held only eight standard frames were popular sev-
ral years ago and have been used by some beekeepers for
pollination, but their use is declining.

The size of the brood nest is the best measure for estimating
the population of the colony. Since the brood nest takes the
shape of a sphere within a colony, one can estimate its total size
by counting the number of frames containing brood. A colony
of bees with brood in five frames will usually have a population in excess of thirty thousand bees. This is considered an
acceptable size unit for pollination purposes.

Colony populations can vary from a few hundred to eighty thousand bees. It is thought that the larger the population, the greater the percentage of bees that will forage from the hive. Colonies of bees that contain less than approximately thirty thousand bees are not efficient because too small a percentage of their force is available for field work. Colonies of bees with an excess of fifty thousand bees would be too crowded if they were contained in only two hive bodies. When colonies of bees become crowded, they may swarm, and swarming should be discouraged since it divides the colony population and both the beekeeper and the renter of the bees lose.

Those who rent bees are at a disadvantage in determining the value of the product they rent. In some parts of the United States quality control is guaranteed by growers who hire bee inspectors to check for colony populations, queenrightness and disease. With some experience one can estimate a colony population by observing the number of bees flying to and from the entrance in a given period of time; however, flight varies with temperature and sunlight. It is a sign of efficient colony management when colonies are equal in population. If nothing else, flight from colonies under similar environmental conditions should be similar.

Package bees from the southern states have been sometimes recommended to pollinate certain crops. Package bees are sold as units containing one, two, three, four or five pounds of bees. There are approximately four thousand bees in a pound. The most common unit sold is the three-pound package, which contains about ten thousand to twelve thousand bees. The number of bees in a package is too few both to protect the brood they start to rear and to forage. Furthermore, the population of the package dwindles as older bees in the package die. Since it takes twenty-one days to produce a bee from egg to adult, in the north it is twenty-two or twenty-three days after a package has been installed before young bees appear in the

Colonies of bees moved into an orchard for pollination in New York State in early spring. Hives are painted different colors to aid the bees in orientation. The colonies have been placed on a pallet so that the bottom boards will remain dry.

The dandelions in this young apple orchard will attract many pollinating insects, including honeybees, away from the apples. Dandelion nectar and pollen may be even more attractive than that produced by apple or other fruit trees. For effective pollination, weeds should be mowed or otherwise removed during blossom time.

hive. During this time there is obviously a considerable drop in population as certain of the older bees in the package die. For these reasons package bees are not recommended for pollination.

The Pollination of Apple and Other Orchard Fruits

More colonies are rented to pollinate apples and orchard fruits in the northeast than for all other crops combined; in fact, probably 15 times more colonies are used for orchard pollination than for all other pollination needs. This is also true in the apple-producing areas of Michigan, Ontario, Washington and Oregon. In California the crops for which thousands of colonies are used for pollination are alfalfa and almonds.

Only a small number of colonies is kept in the vicinity of orchards since they are usually not good honey-producing areas and also there are problems from pesticides throughout the year. This means that colonies must be moved to the orchards, which poses special problems for the beekeeper. Fruit trees bloom in May in most of the north. This is a time of year when the colonies have not yet reached their maximum strength, and when the brood nests are still expanding and the colony population growing. Colonies in the process of expanding their brood nests and increasing their populations have special needs and can easily be placed under stress if not given the care and attention required. At this time of the year the control of temperature and humidity and the need for food in the colony on a day-to-day basis are both critical and difficult because of inclement weather. This is a time of year when any stress can result in an outbreak of nosema, European foulbrood or sacbrood, all of which are chronic in honeybee colonies.

Colonies used for pollination in the spring should be in two standard Langstroth supers (two-story colonies) and contain about thirty thousand bees and a queen. Though beekeepers do not care to carry any more weight than necessary, such a colony should have a minimum of fifteen pounds of honey when it is taken to the orchard. Experience in orchards in the early spring indicates that inclement weather can prevent flight for several days, and colonies with any less than this quantity of honey may, in an occasional year, be unable to sustain themselves. Most colonies used for apple pollination are usually widely scattered and it is not feasible to try to feed them while they are in the orchards.

Colonies of honeybees in an orchard should be placed on some kind of hive stand or on stones or a pallet, something that will raise the bottom board two to six inches off the ground and keep it dry. When moisture accumulates within a colony or when a bottom board becomes wet, the process of evaporation cools the hive unduly and places a stress on the colony. Colonies should be located in full sunlight so as to facilitate keeping them dry and so that bees will take flight as early as possible in the morning. Where possible, colonies should be located near a source of good clean water. Bees use large quantities of water in the spring to dilute the food fed the young larvae in the hive. If the only source of water is contaminated water from wheel ruts and ditches in the orchard, there may be some insecticide poisoning of colonies.

In actual practice in the north about one colony is used for three or four acres of fruit. Probably a greater concentration of bees would be desirable, but most growers are not convinced of this. It is recommended that colonies be located in groups of three to five, to facilitate both loading and unloading, but more especially to take advantage of special climatic conditions and available sunlight in the orchard. Bee flight is controlled by temperature and sunlight. Colonies that receive maximum sunlight and are located in the most favorable locations will send the greatest number of workers to the field for pollination.

The Pollination of Other Crops

In the North a small number of colonies, four thousand to six thousand, are rented to pollinate bird'sfoot trefoil, cucumbers, squash, cranberries, blueberries, cantaloupes and other crops. In Maine, New Jersey and Michigan thousands of colonies are rented to pollinate blueberries. Cranberries in Massachusetts and New Jersey is also a crop for which bees are rented. In certain areas in the United States small or amateur beekeepers can earn additional income by renting their bees for pollination. Often, when a grower needs only ten to twenty colonies, he is willing to provide a truck in return for renting the bees for a lower price. There is an added advantage for a grower who rents only a few colonies and trucks them to and from his farm himself—under these circumstances colonies can be taken to the field or orchard at the last minute and the bees will work close to the hive and on the crop where they are wanted.

While the problems vary from one area to another, the number of colonies per acre, the size of the colony population and the placement of colonies are not very much different from those used in the pollination of orchard crops. County agents and state apiculturists should be consulted for information on specific pollination problems.

XIII

The Biology
of the Honeybee

The honeybee is not a domesticated animal. It is possible for man to keep bees in a hive only because he understands their biology. Beekeeping is the application of the knowledge of bee behavior.

Progress has been made in bee breeding in recent years, but it is a slow process. Men have selected certain strains of bees for their effectiveness as honey gatherers and pollinators, their ability to resist disease, their gentleness and their adaptation to a wide range of climates. A method of artificial insemination has been developed that has been helpful in further exploration. As our knowledge of basic bee biology increases, we should be able to make changes that will benefit both the commercial beekeeper and the hobbyist.

Each caste in the honeybee colony is highly specialized

and dependent upon another caste. The queen is an egg-laying machine; additionally she gives the colony certain chemicals important for the maintenance of social order. The sole function of the drone is to mate. To the worker honeybee falls all other tasks. However, the adaptability of the worker and her ability to change from one task to another with only a few seconds' notice are remarkable. Such adaptability is required for the colony to survive during times of stress.

Because the evolutionary pressures on the three castes differ, it is reasonable that their life histories should vary and that even their development time while in the egg, larva and pupal stages should not be the same. Table I indicates that the queen develops from an egg to adult in the shortest time while the drone takes the longest time. The queen is the most important individual in the honeybee colony. From an evolutionary standpoint it is reasonable that her development time is less than that required for the other two castes, since rapid queen replacement may have much to do with the survival of a colony.

Because the sole function of the drone is to mate, and drones are easily reared during the mating season, there has been no evolutionary pressure to speed up their development time.

The number of days required for the development of each caste as listed in Table I are reasonably accurate; however, queens for example, have been known to develop in 15½ days. There is some research to suggest that during the cooler months of the year worker development may actually be slowed because the bees cannot maintain the brood-rearing temperature with ease. However, unlike most insects whose development time may vary greatly because of fluctuating temperatures, the uniform temperature in the bee hive means that brood may develop in set periods of time. This fact unquestionably works to the advantage of the honeybee in its competition with other animals.

TABLE I

Honeybee Caste Development Time in Days

Caste	Egg	Larva	Pupa	Total Time
Queen	3	5½	7½	16
Worker	3	6	12	21
Drone	3	6½	14½	24

Life History of the Worker

The worker bee is a female; she arises from a fertilized egg. The same egg may produce a worker or a queen depending upon the food received during early larval life. In general, it may be said that a worker honeybee lives for six weeks in the summer and for six months in the winter. Like most general statements about animals, it must be understood that there are a great number of variables. In the winter the worker works less than in the summer, or at least the work done is less wearing on the body. The work done in the summer ages the worker much more rapidly. Old bees can be identified by their physical appearance. They have less body hair and their wings may be frayed. Additionally, adult bee diseases may shorten the life of worker bees. Nosema and paralysis and many lesser maladies play important roles in this regard. Adult bee diseases are especially a problem when colonies are under stress.

In the active season a worker honeybee undertakes a long series of duties after her emergence from her cell. These usually follow a set pattern; however, one of the marvels of the hive is the fact that a worker bee may change occupations, sometimes within minutes, if there is a real need in the colony. For example, if there is a sudden need for a large quantity of water to cool the hive, bees will be diverted from other tasks to do this job. Likewise, many bees may engage in ventilating—fanning their wings to move volumes of air through the hive both to evaporate water to cool the hive and to evaporate excess moisture from freshly gathered nectar. Another example of the bees' ability to change jobs rapidly is seen when a large number of bees are required to defend the colony.

When a bee emerges from her cell, she first engorges on pollen and nectar. Following this, she engages in cleaning cells, perhaps even the one from which she herself emerged. This is in preparation for further egg laying by the queen or in anticipation of cells needed for honey storage. The next activity is that of feeding honey and pollen to the older larvae within the hive. This is followed by feeding the younger larvae; since young larvae receive a specially secreted food from glands in the heads of worker bees, these glands must be given time to develop after emergence of the worker from her cell and so there is some delay before this particular task is undertaken. The number of days spent feeding brood depends upon the quantity of brood to be fed and also the pressure for other tasks to be undertaken within the hive.

Following brood feeding, worker bees may engage in comb building and cell capping (the capping of cells with larvae or those with honey). At this time the worker may also engage in taking food from the field bees. Field bees do not place the nectar they collect directly into cells but rather give it to house bees, which manipulate it further, adding an enzyme and aiding in the reduction of moisture. The last task undertaken by a house bee before she becomes a field bee is that of guarding the hive. A worker bee is usually eighteen to twenty-one days of age when this occurs. At this time her sting glands contain the maximum amount of venom, so she is an ideal guard. A worker bee remains a guard usually less than a day or two.

At about three weeks of age a house bee becomes a field bee. She will not again undertake household tasks unless there is a drastic need for her to do so; however, research shows that she retains the ability to do several tasks, except probably that of secreting special food from her head glands, which have probably atrophied by this time. It is also questionable if old bees can again secrete wax. As a field bee, the worker engages in the collection of nectar, pollen, water or propolis, again depending upon the needs within the colony. At this time in her life the worker bee undergoes the greatest stress. Infrequently a bee will remain out all evening if she happens to be working late at night and is chilled; this can place a great stress on her body. Similarly, winds can fray and tear her wings. There are always other insects and birds that prey upon her as she moves from one flower to another.

Work within the colony never ceases. There is never a time when the whole colony sleeps, but as one watches individual bees, it is evident that they do rest periodically throughout the day. This is especially true of house bees and less true of field bees; presumably the field bees rest most of the night. Furthermore, it will be noted that a single bee spends a great deal of time walking over the comb as though she is looking for work to be done. For example, rarely does one see a worker bee completely cap a cell of honey or a brood cell by herself. Rather, the capping of a single cell is done by several bees that happen upon it and recognize it as something that must be done. It is presumed that this apparently random movement of bees around and through the hive is part of the system that alerts the bees to the work to be done.

Life History of the Drone

Many people have searched in vain to find some activity other than mating for which the drone can be credited. It has been suggested that a large number of drones on the outside of a cluster in the spring helps to insulate it and protect the brood, but if this is so—and it is unlikely—the function is accidental. The drone's whole body is adapted for his sole function. He has no sting or any of the other special body modifications that aid the worker in her role.

Drones are not present in the honeybee colony in the winter. Drone rearing starts in the spring about the time pollen and nectar become available in large quantity and when the rearing of drones would place no stress on the colony. Similarly, drones cease to be fed in the fall and finally, when they are starved and weakened, they are dragged from the hive by worker bees. Contrary to popular notions, not all of the drones are driven from the hive at the same time in the fall. Rather, it appears that the older drones die first, and a very small number of drones persist even after the first frost.

Drones become sexually mature at about ten to twelve days of age. At this time they begin to take short orientation flights and later the longer flights required for searching for their mates. Both virgin queens and drones fly only in the afternoon, thus increasing the chances that the sexes meet for mating. Drones, and presumably virgin queens, fly to what are called "drone congregation areas," where mating takes place. Drone congregation areas are well defined in hilly regions and such areas may be only a few hundred feet in diameter. In flat areas drones may range over larger areas.

A queen mates with several drones. Drones die in the mating process. Proportionally to its body size, the genitalia of a drone are among the very largest of any animal on earth. They are contained in the abdomen and presumably getting them out of the abdomen for the purpose of mating places such a strain on it that it dies in the process.

Young drones feed themselves from honey cells within the colony. However, drones soon learn to solicit food from workers and become very adept at doing so. Old drones, even when they are starving in the fall, have apparently lost the ability to feed themselves and continue to solicit food from workers even though they are unsuccessful in doing so.

Perhaps most interesting is the fact that no one has ever captured a queen honeybee outside of a hive on a flower. There are only three records of anyone ever capturing drone honeybees on flowers outside of the hive. In all three cases the drones were found on goldenrod flowers in late fall and were presumably old, starving drones. Their being on goldenrod flowers was probably accidental and there is no indication that they were feeding. It is an interesting fact that queens and drones feed only within the colony and apparently never in their normal life alight to rest or feed in the field. This means, too, that queen and drone flights are of limited duration, since more food is required for flying. Queen and drone flights average about 20 minutes or, rarely, twice that long.

Life History of the Queen

A queen honeybee is a very special animal. A queen is important to the colony for two reasons: she lays all the eggs and she produces secretions from glands in her body that are responsible for the maintenance of social order.

Biologists interested in development and nutrition have pointed to the queen as an example of how food may make tremendous differences in an animal. Queens are fed lavishly by worker bees as witnessed by the tremendous amount of royal jelly—a thick, white, creamy material found in the base of the cell in which a queen is reared. Queens are markedly different from worker bees. Some of the more apparent morphological differences include the fact that the queen has no wax glands, her sting is lightly barbed and curved, and she has no pollen baskets on her hind legs or other modifications on the forelegs as does the worker. While she has certain of the head glands found in worker bees, these secrete radically different

Bees in a flying swarm are aware of the presence or absence of their queen; under special circumstances one may lead a swarm for long distances with a caged queen as is being done in this illustration. Beekeepers with a thorough knowledge of swarm behavior use the same technique to build a beard of bees. These bees would soon settle on the beekeeper's hand if he stopped walking.

materials. Perhaps most important, the ovaries of the queen are much better developed than are those of the worker bee; a queen has a spermatheca, in which sperm is stored for long periods of time, even years. A worker bee is incapable of mating or storing sperm.

Queen honeybees have been known to live five or six years, though in commercial practice, colonies are usually requeened after one or two years. The long life of the queen is important for the continued existence of the colony, and natural queen replacement is a delicate operation whose failure can cause the loss of the colony.

Queens are reared only when a colony is about to swarm, when a queen is old and failing and is superseded and when, for some reason, the colony becomes queenless. When a colony becomes queenless, the honeybees detect the loss of their queen, usually within a few hours, and immediately begin to enlarge the cell around an egg or one-day-old worker larva in the colony, and begin to construct a special queen cell in which to rear a new queen.

When a queen first emerges from her cell, she is not detected by the workers as being a queen. The workers pay little attention to her; it is several days before she begins to produce her queen secretions. A virgin queen, upon emergence from her cell, engorges on pollen and nectar as does a worker bee. Following this, the queen seeks out other virgin queens and fights and attempts to kill them. If virgin queens are not present, she turns her attention to other queen cells and destroys them. A queen honeybee leaves the colony on only two occasions—when she is mating and when she accompanies a swarm. Mating takes place when the queen is five or six days old. About half of the queens mate on their first flight, but some take an orientation flight prior to mating. Queens mate with an average of about eight drones in a matter of one or two days and start to lay eggs two or three days later. From these several matings a queen obtains in the vicinity of five million sperm, which will last her the rest of her life. Queens are incapable of mating after they are about three or four weeks old, and old queens that have exhausted their sperm produce eggs that develop only into drones.

Within the colony, the queen is always surrounded by a few worker bees who lick and groom her and feed her. A single worker usually remains with the queen only a few minutes.

Knowing when a queen should be replaced is important in practical beekeeping. Old queens have fewer body hairs and a black, shiny appearance. The number of worker bees surrounding an old queen is smaller than the number surrounding a young queen, because the old queen is not so attractive to the workers in her hive. However, the easiest way to determine a queen's effectiveness is to check her brood pattern. A young, vigorous queen lays eggs in a compact area in the brood nest. As the queen grows older, more and more cells in the brood area will be left without eggs. Brood in adjacent cells should be of a similar age; if it is not, the queen is not laying in all cells or her eggs are not hatching.

Life History of the Colony

Colonies of honeybees undergo what beekeepers term a "cycle of the year." In all parts of the northern hemisphere the queen lays the least amount of eggs in October and November. It is normal for colonies in Florida and even the far north to start to rear brood in late December and January; however, the quantity will vary because of temperature limitations. This coincides with the time that the day length is increasing and it has been suggested that an increasing day length is a factor that controls egg laying by the queen and brings about an increased population in the spring. In this regard it is interesting that when the days start to shorten in late June, the colony division or swarming becomes less of a problem. However, this is only a theory and final proof to demonstrate the factors that control egg laying and colony development are wanting.

In the southernmost states colony populations can begin to increase in late January as a result of the egg laying in December and early January. In the more northern states colonies reach their low population of the year in late January and February. In the northern states the egg-laying rate is not sufficient to keep pace with the death rate at this time of the year.

Across the southern states colonies reach their peak populations in March, while in the north the peak populations are reached in late June and July. In the northern states it is not uncommon for colonies to have a low population of ten thousand to fifteen thousand bees in February and a peak population of eighty thousand bees in July. In the northern states colony populations may drop in late July and show a slight population increase again in August when goldenrod and other late summer and fall plants provide the pollen and nectar to stimulate brood rearing.

It is because colonies of honeybees show these seasonal variations that many people have suggested we should consider the whole colony as a single animal. This has considerable merit since none of the individuals in the colony can live alone and there must be a sharing of work for the colony to survive. Furthermore, in the spring the colony shows an increase in population or growth that is somewhat akin to the growth of other animals. Concomitantly, when the colony suffers from disease, usually all the members of the colony are affected and one cannot treat individuals separately in most instances. It is important for every beekeeper to determine when changes take place in colonies in his area. These changes are determined by the time of year, the weather and the food available to the colony. Once the cycle of the year is thoroughly understood for a specific area, the beekeeper is in a position to prepare a sound management program.

Communication and the Senses

When many individuals live together, they must have some method whereby they can communicate with one another. Honeybees in a colony face many of the same problems humans face in their society. They must be able to recognize one another and also their queen. They must be able to alert one another to danger. When food is found in the field, it is important to tell other members of their community where it is. Within the hive tasks must be done in an orderly fashion. Thus, the need for an efficient communication system is evident.

In the case of man, speech is the most important form of communication. Visual cues, physical movements and odors are important but much less so than talking. Honeybees have a very acute sense of smell, far better than that of man. As for taste, honeybees can taste about the same things that man can taste insofar as sweet and sour, bitter and salt are concerned. Color is also important to the honeybee, though the bee does not see so well as man.

Honeybees use chemical substances, especially as odors, to give information to one another. Chemical substances that convey messages from one member of a community to another are called "pheromones." This term was coined in the late 1950s. In precise terminology a pheromone is a chemical substance secreted from a gland in one animal that brings about a specific response in another of the same species. So far as is known, honeybees do not use sound as a method of communication and no sound receptors are found anywhere on their bodies; however, honeybees do detect vibrations and respond to them, presumably through organs on their feet. Such vibrations are sometimes referred to in the literature as being "substrate-born sound."

In the case of the honeybee, several pheromones have been discovered. The role that certain of these play is known and is discussed below. Honeybees use their antennae to detect odors, and microscopic examination of the honeybee's antennae reveals that they are covered with a great number and variety of sensory receptors.

Sight

Honeybees see four colors distinctly: yellow, blue-green, blue and ultraviolet. Man can distinguish about 60 colors distinctly and several shades under certain circumstances. Red, on one end of the color spectrum, is seen as black by bees; on the other end of the spectrum bees see ultraviolet, which we cannot see.

Bees do not see designs so well as does man. For example, a bee confuses a two-inch circle and a two-inch square of the same color, but they are able to tell the difference between a solid circle of one color and a circle with only an outline of the same color.

Experiments to show the colors and designs honeybees see were performed by Professor Karl von Frisch several years ago and are simple to duplicate. One of Professor von Frisch's experiments involved training bees to go to a feeding station in the field. The feeding station can be placed on a small table painted blue or that has a blue piece of cardboard immediately under the feeding station. After scout bees had made several round trips from the hive to the feeding station, von Frisch would offer the bees a choice of two, three or four stations close together but only one of which was marked with blue. Under these circumstances most of the bees alighted at the feeding station that carried the proper color and did so without hesitation. Another experiment involved training bees to feed at a station inside a colored box with only a small hole leading to the outside. Again, after a suitable training period, the bees would be offered two, three or four boxes of similar size but colored differently. And again, in this experiment the bees entered the box marked with the color to which they had been trained. The same experiments may be used to show the shapes and designs bees see distinctly.

Color and design are obviously important in the life of the bee. Once a bee finds a good source of food in the field, it is important that she be able to move from a flower of one kind to another flower of the same kind. The bee does so by using the color and design of the flower. Odor is also important and is discussed below. Insofar as practical beekeeping is concerned, and as is mentioned elsewhere in the text, painting bee hives that are in close proximity to one another different colors helps bees distinguish their own hive from others in the apiary.

Odor

As he had done with color and design, Professor von Frisch investigated the ability of the worker honeybee to detect odors. The design was simple, and the experiments are easily repeated and demonstrate very clearly that honeybees detect odors.

Professor von Frisch placed scented foods in boxes with only a small entrance. When the bees had been trained to the scented box, they were offered a choice of several boxes, but only one contained the proper odor; the bees entered the proper box without hesitation. It was shown further that honeybees detect odors with their antennae. If a terminal eight segments are removed from one antenna and a terminal seven segments from the second, the bee can still detect an odor. If a terminal eight segments are removed from both antennae, the bee is unable to detect odors, indicating that the most important sensory receptors lie on the terminal eight segments.

It was logical to ask next which might be more important, color or odor. When Professor von Frisch trained bees to feeding stations on the bases of both color and odor, he learned that the bees used color from a great distance, but failed to enter the experimental feeding station or to feed if the food was not properly scented.

Taste

As indicated, honeybees can detect sweet, sour, salt and bitter. If, for example, salt is added to a sweet sugar syrup in too great quantity, the bees will reject the syrup. Interestingly, however, honeybees cannot detect all the tastes that man can. For example, bees do not taste quinine and one may add quinine to a sugar syrup, which will be extremely bitter to man but bees continue to feed on it without hesitation.

Honeybees have a threshold of perception and also a threshold of acceptance. They are able to detect very low sugar concentrations, but they will feed on them only when there is nothing else available. Thus, in the field scout bees have a tendency to work the richer sources of nectar. As a practical example, one may point to the problem of pollinating pears in the United States. Pears produce nectar that contains only 10 to 15 percent sugar while apples, dandelions and other weeds in flower at the same time usually have a richer sugar concentration in their nectar. And as a result, growers do various things, including eliminating competition, in order to have their pears properly pollinated.

Pheromones

The idea that chemical substances exchanged by individuals within the colony might have something to do with social order was postulated in the 1940's and 1950's; however, it was not until 1961 that the first of these chemical substances was isolated and synthesized so that controlled laboratory experiments could be made. That year Dr. C. G. Butler of the Rothamsted Experiment Station in England and his colleagues identified queen substance. In the colony this material is said to have three functions. Queen substance is the material by which honeybees recognize their queen; it inhibits queen replacement and so long as the queen is present producing her secretions, the bees will not build queen cells; and lastly, it is thought that queen substance inhibits ovary development in worker bees. The source of queen substance is the queen's mandibular glands. However, when the mandibular glands of a living queen are removed, the worker bees within the colony are still able to recognize her, though to a much lesser extent. Therefore, it is clear that secondary substances also play roles in queen recognition, etc. But these materials have not yet been identified chemically.

Queen substance is the material that causes honeybees to surround and lick their queen. One can place queen substance (synthetic or natural) on a piece of wood or other object, and observe that worker bees will surround it, lick it and attend it in the same way that they will their queen. The fact that honeybees do this clearly demonstrates that they are not thinking animals but merely responding to stimuli in their environment.

In 1962 it was discovered that queen substance was also the honeybee sex attractant. It is an interesting fact that inside the hive drones pay no attention to queens; mating always takes place outside of the hive and usually at a height greater than 20 feet. Again, experimentally one can use the synthetic queen substance to attract male honeybees by placing it on a wad of cotton, piece of wood or other object and suspending it from a helium-filled balloon. When this is done, there is clear evidence that the male honeybees, too, are responding to an odor only and that they do not recognize a queen as a living animal.

The fact that one chemical substance can play so many roles in a honeybee colony is perhaps not surprising. In the course of evolution animals have usually found the most efficient way of communicating and maintaining social order.

Isoamyl acetate, the alarm odor in honeybees, was discovered in 1962. This chemical substance, secreted from the vicinity of the worker honeybee's sting, alerts other bees to danger and causes them to attack. It is interesting that the honeybee alarm odor is a common chemical that has been on laboratory shelves for years, but no one had ever thought of testing it as an alarm odor. As with queen substance, a very small quantity of the synthetic material dropped in the vicinity of the hive will cause bees to attack and sting.

It is also interesting that honeybees will attack as a result of the release of alarm odor only in the vicinity of their hive. Perhaps this is logical, for it is the hive they must protect. If alarm odor is released in the vicinity of a worker bee in the field, she will flee, but in the field there is nothing for her to defend except herself.

In 1966 it was found that there is a second alarm substance in honeybees which is secreted from the worker's mandibular glands. Presumably it is with this material that worker honeybees may also mark an enemy. In any event, it is no accident that honeybees are able to follow an enemy. So far as we can determine, the reason that a sting is left in a victim by a honeybee is for the precise purpose of marking the enemy, as the alarm odor will continue to be released no matter where the enemy flees.

Worker honeybees mark food sources and, under certain circumstances, their own hive or a new nest by exposing the scent gland on the tip of their abdomen. Four chemicals are released at the same time from this gland. The identity of these was discovered and reported in papers published in 1962, 1964 and 1966. As might be imagined, an individual scent gland on a worker honeybee secretes a very small quantity of scent and the identification of these materials involved a remarkable piece of chemical sleuthing.

From this research it is clear that there are many pheromones that control other aspects of social order in the honeybee colony. Investigators in many parts of the world are exploring the honeybee community in order to identify these substances and the roles they play. Since the quantities of the materials involved are small, the research is slow and tedious and demands the use of very sophisticated chemical apparatus.

The Dance Language

Perhaps the most fascinating aspects of honeybee biology are the dances worker bees perform to tell other bees in the hive about food sources in the field. There are basically two types of dance, the round dance and the wag-tail dance, both involving physical movements and conveying very definite messages. What is most intriguing about the dances of the honeybee is that man can read them, but we do not know how honeybees do so. Research published in 1971 shows that worker honeybees must follow several dancing bees in order to perceive the message conveyed in the dance. Not all of the bees that follow dancing bees and leave the hive necessarily find the food source.

Research on the dance language of the honeybee has been undertaken almost exclusively by one man and his students, Professor Karl von Frisch of Germany. Professor von Frisch began working with honeybees as early as 1914, but it was not until the late 1940s that he came to understand the honeybee dances. The basic von Frisch experiments have been repeated by many people and are now routinely used in laboratories to demonstrate one form of animal communication.

More than physical movements alone are used to convey information about food sources in the field to bees that are being recruited by scouts. A scout bee that has found a rich source of nectar will also pause frequently and give a taste of the nectar to a potential recruit. In doing so, the recruit is able to perceive the sugar concentration of the food and also to determine its odor. If the food source is close to the hive, odor from the flowers in which the bees have worked in collecting their nectar clings to the outside of their bodies and is detected by recruits.

Precisely the same dances are used to convey information about pollen and, to a much lesser extent, propolis and water when these items are needed in the hive. While it is clear how scout bees convey information that the item to be collected is nectar or water by giving a taste of it to recruit bees, it is not at all clear how they convey information about the fact that it is pollen or propolis. This mystery remains to be researched by some thorough, careful investigator. When honeybee colonies swarm, the same dances are used to indicate a new home site. Again, how it is told that it is a home site that is being danced about is not at all clear; what is clear is that the proper message is conveyed to the bees in a swarm, for experimentally it can be shown that they move to the new site without hesitation.

The simplest of the honeybee dances is the round dance. The round dance merely indicates to recruit bees that food is to be found within about a hundred yards of the hive and that they should go out and search for it. Von Frisch proved this point by setting up a feeding station within a hundred yards of a hive and trained a certain number of scouts to go to the feeding station. These were marked so that they could later be differentiated from the recruits that came to the feeding station. After a given training period, the original station was removed and four feeding stations were put up at cardinal points and equal distances from the hive. If the bees that visited these were counted, it could be noted that the bees came in about equal numbers to all four feeding stations. This was not the case when scouts danced the wag-tail dance.

Certain races of bees use a dance intermediate between the round dance and the wag-tail dance incorporating parts of both. Such a dance is used to indicate distances of usually fifty to about a hundred yards. This intermediate dance was not discovered by Professor von Frisch, for the race of bees he was using experimentally did not perform this dance. This demonstrates the kind of biological variation that can exist among animals and serves to warn all investigators of the kind of error that can be made.

Perhaps the most interesting of the honeybee dances is the wag-tail dance, which conveys to recruits information about the direction and distance of the food source from the hive. The dance is remarkably simple and may be followed by anyone using an observation hive. The fact that so much information can be conveyed by a dance is remarkable.

The rapidity with which the dance is done by a scout bee indicates the distance of the food source from the hive. For experimental purposes, Professor von Frisch chose to count the number of completed dances within a 15-second period. He placed these on a graph and showed that when food sources were close to the hive, the bees were more excited and danced more rapidly. When food sources were at greater distances the scouts danced more slowly. Just as man himself may show enthusiasm by the rapidity of his movements or speech, so, apparently, honeybees use the "enthusiasm" of their dance to indicate the closeness of a food source. It is thus clear that if two bees are dancing side by side and one indicates a food source close to the hive, while another indicates a food source farther away, recruits will be more attracted to the bee dancing more rapidly.

The inside of the hive is dark. The part of the dance made

when the bee is running the wag-tail itself indicates the direction of the food from the hive as oriented by the sun. In the hive, the worker honeybee transforms sun direction into gravity. Therefore, if a bee dances directly up on a comb, she is indicating that the source of the food is in the direction of the sun. If the same bee dances directly down, she is indicating to recruits "go away from the sun" to find food.

Again, in what has become a classic experiment, Professor von Frisch showed the accuracy of the wag-tail dance with ease. After training bees to a food source in the field for a given period of time, he established a series of feeding stations on both sides of the original feeding station and then began to count the number of recruits that appeared at all stations. He found that a remarkably large number of bees arrived at the right station but that a certain number of bees came to feeding stations on either side of the station to which the scouts had been trained. The fact that bees can find the feeding station under these circumstances is remarkable, yet it is clear why there might be variation. Under normal circumstances a scout bee dances to indicate a clump of flowers or a tree from which recruits might gather food, or, more commonly, to indicate a field or a forest in flower. In fact, were there not an abundance of food in the field, the scout bee would not be sufficiently enthused to dance at all.

One other factor that influences the enthusiasm with which scout bees dance is the way in which their food is received in the hive. Field bees do not deposit nectar in cells themselves but rather give it to house bees. If a scout bee is not able to find a house bee to take her food because they are all busy, a scout bee is much less inclined to dance and dances for a shorter period of time. The same is true with pollen. If the scout bee cannot find cells within the hive in which to deposit pollen, she is much less inclined to dance to indicate a pollen source.

XIV

Honey Wine

Honey wine, better known to some as mead or metheglin, was probably the first alcoholic beverage made by man. Honey wine is made by diluting honey with water and adding yeast. Yeasts, which are living cells, ingest the sugar in honey and convert it into carbon dioxide and alcohol. Honey wines contain about the same amount of alcohol as grape table wines, usually about 12 percent.

Honey wines may be light or dark, depending primarily on the color of the honey used. Many honey wines are often sauterne-like in physical appearance.

History

History reveals that all the ancient civilizations, including the Egyptians, Greek, and Roman, made honey wine. Beekeeping was popular with all these peoples, and parts of the area around the Mediterranean Sea, even today, are known for the honeys they produce.

In North Europe there is an especially rich history concerning honey wine. The early Scandinavians, who presumably drank from the skulls of their slain enemies, made mead. When the Roman Legions first invaded England they found the people there making alcoholic beverages from honey and apple juice. Still later in English history honey wine became the national drink and both mead makers and beekeepers were often members of the official court.

Honey winemaking was popular in Poland and Germany. People in these areas practiced an elaborate type of forest beekeeping, harvesting their honey from bee trees. In many parts of North Europe there was little other than honey from which man could make an alcoholic beverage.

Types of Honey Wine

There are three basic types of honey wine. Most popular is a standard mead in which one dilutes the honey with water, adds a small amount of nutrients and ferments the mixture in the normal fashion. Such a mead depends upon the flavor of the original honey for the flavor of the final product. A second type of honey wine may be made by adding a fruit juice to the honey-water mixture. Fruit juice substitutes for the nutrients in the first formula. Honey fruit wines depend upon both the honey and the fruit juice for the flavor of the wine. A third type of honey wine is known as metheglin, a spiced honey wine. Wines of this sort were especially popular in early England and many of the old recipes have survived. The spices that are commonly added to metheglins include cinnamon, ginger, and nutmeg.

Sparkling honey wines, like sparkling grape beverages, have been produced in many parts of the world. It is possible to make any of the three basic types of honey wines into a sparkling mead. First make a standard honey wine, but one that is dry (that is, one without sugar) and contains only about 10 percent alcohol. Then place the fermented beverage in a champagne-type bottle and add more sugar and yeast. A second fermentation takes place in the bottle. The second fermentation produces an additional amount of alcohol plus carbon dioxide, which gives the sparkling beverage its bubbles.

Fermentation Vessels

Traditionally, wines have been made in oak or wooden casks. In part this has been true because there have been few other satisfactory vessels for holding a fermenting beverage. Today, in commercial winemaking, vintners are increasingly using stainless steel, epoxy-lined, glass-lined and concrete tanks for their fermentations. The reason is simply that such vessels are more easily cleaned and sanitation is less of a problem.

Five- and ten-gallon glass carboys make excellent fermentation vessels for home winemakers. One-gallon glass jugs are satisfactory for experimentation but the quantity of wine or mead produced is so small that their use is impractical. Glass carboys have two distinct advantages insofar as the winemaker is concerned: such vessels are easy to keep clean and one may follow the course of the fermentation, observing both the fermentation itself and the clearing of the new wine.

Sanitation

A variety of microbes may spoil the wine either while the initial formentation is underway or after it is finished. It is important that the home winemaker keep all his equipment scrupulously clean and that he use only good yeast cultures to bring about the fermentation.

Active yeast cells can live in the presence of relatively high levels of sulfur dioxide. Since most microorganisms, especially those that might harm a wine, cannot tolerate this chemical, nearly all commercial wineries add sulfur dioxide to new juice prior to the fermentation and again when the new wine is bottled. Home winemakers should do likewise. The home wine shops catering to the needs of amateur winemakers sell pre-measured tablets that may be added to a fermenting wine so as to release the proper amount of sulfur dioxide.

A one-gallon jug with a fermentation valve is used for this mead fermentation; a homemade and an alternate type commercial fermentation valve are shown in front of the jug. Note the foam on the fermenting mead caused by the escaping carbon dioxide.

Commercial wineries prefer to use soda ash to wash their barrels and equipment. A chlorine bleach solution or soap might be equally good but these are often toxic to yeasts and may deter their growth if they accidentally contaminate the liquid to be fermented. However, one may use them to wash the walls, floors, or a laboratory bench where honey wines are made, thus deterring the growth of microorganisms in these areas.

Fermentation Valves

During the course of a natural fermentation a large volume of carbon dioxide is given off. This must be allowed to escape, yet, at the same time, one must keep the fermentation vessel closed so that harmful microbes cannot enter. One method is to plug the mouth of the carboy with clean, dry cotton. The carbon dioxide will move through the cotton with no difficulty, but harmful microbes which are airborne cannot do so.

Fermentation valves, which are usually water valves and allow the pass-through of carbon dioxide, are popular with home winemakers for several reasons. They are cheap and easy to install. More important, as the carbon dioxide bubbles through the water in the fermentation valves, one becomes aware of the rapidity with which the fermentation is proceeding. When the fermentation is completed the fact is made obvious by the clearing of the new wine and also by the slowing of the passage of carbon dioxide through the fermentation valve.

Selecting the Yeast

Yeasts that are used to make any alcoholic beverage belong to the same genus; however, hundreds of strains or races exist. Commercial vintners select their yeasts with great care, for the yeast can have a profound effect on the flavor of the final product. Yeasts used to make breads and cakes are not satisfactory. If they are used the final product will have a "bready" flavor.

Certain yeasts have been preferred for the express purpose of making mead. However, in the author's experience, any yeast that makes a good, light grape wine is satisfactory for honey wine fermentation. Pure yeast cultures are available from several sources including many of the wine shops and supply houses for home winemakers.

The directions on most yeast cultures call for adding an active yeast culture to the medium to be fermented at the rate of about 10 percent by volume. A broth or mead containing an active yeast culture will release carbon dioxide and form a generous foam when the culture is swirled gently. The aim is to introduce into the new juice or mead a sufficient number of yeast cells so that fermentation is rapid and overcomes any other microorganisms that might be present. Some home winemakers add as little as 2 to 5 percent of an active culture by volume; such a small amount may work but it is risky to use so little yeast.

Formulas

Yeast cells are living organisms. To produce carbon dioxide and alcohol they must grow and flourish. To do so they must be properly fed. While sugar is the chief food of the yeast cell, yeasts, like other living plants and animals, need certain vitamins and minerals.

Honey contains only small quantities of materials other than sugar and water. When the honey is diluted to make a honey wine the vitamins and minerals in it are also diluted. It is not practical to dilute honey and leave it to ferment naturally without the addition of yeast food. Furthermore, the dilution of honey to make honey wine also dilutes the acid present in the honey. It is therefore important that additional

At the end of the fermentation, dead yeast cells settle to the bottom of the container forming a white ring; this is part of the clearing and aging process.

acid be added to the mixture prior to the fermentation.

The deficiencies of diluted honey, both in terms of yeast food and acid, may be corrected by adding either pure chemicals or fruit juices. Both methods are satisfactory though the choice will have a profound effect on the final flavor.

The basic formula for honey wine is as follows:

4 pounds honey
1 gallon water
4 grams citric or tartaric or malic acid
   (or blend of all three)
4 grams ammonium phosphate
4 grams cream of tartar

The above items are blended and yeast added. One may use crystallized honey to make mead provided it is first dissolved in the water. It may be helpful to warm the water slightly to dissolve crystallized honey. A few beekeepers add a small amount of lemon juice to the above formula; the juice of one-eighth to one-half a normal lemon is sufficient. This adds additional acid and a small amount of nutrients.

One may substitute fruit juice for the above chemicals. The quantity of fruit juice used depends upon the individual's taste and the quantity of acid in the fruit juice. Although juice from wild grape or elderberry is very high in tannic acid, an excellent fruit honey wine may be made using these fruits. Most beekeepers who make a honey wine using elderberries or wild grapes add 10 to 50 percent fruit juice by volume. If only 10 percent fruit juice is added, one need add little or no extra honey to the mixture; however, if one adds as much as 50 percent juice from one of these fruits, then considerable honey must be added since the juice of both fruits is low in sugar; the correct amount is two to three pounds of honey per gallon of juice.

It is desirable to have about 22 percent sugar in the medium to be fermented. Elderberries and wild grapes usually have only 5 to 8 percent sugar content; the mead maker must calculate the amount of honey to be added accordingly. Also more aging is required if greater quantities of fruit juice are added.

The most popular fruit juice used to make mead is apple; second in popularity is pear juice or pear cider. Both apples and pears contain 10 to 12 percent sugar in their juice. Depending upon the quantity of juice added, one should add honey to make up for this deficiency in sugar. The amount added is usually one and one-half to two pounds of honey per gallon of juice. One may make an apple-honey wine merely by adding two pounds of honey to a gallon of cider and allowing the mixture to ferment. If fresh, unpasteurized cider is used there will be a sufficient number of yeast cells present for the fermentation to start; try to avoid cider that contains a preservative. If you must use pasteurized or filtered cider, you will have to add a yeast culture. (The honey must be thoroughly dissolved in the cider before the fermentation is started.)

Metheglin, or spiced mead, is becoming increasingly popular among beekeepers and mead makers. The flavor of a spiced mead is not derived from the honey alone but it is important that a honey with a good flavor be used. The spices used should not be allowed to steep in the water, or honey-water mixture, too long. Otherwise a bitter taste may result. Steeping most herbs for twelve to twenty-four hours appear to be satisfactory.

The most popular spices for metheglins are cinnamon, nutmeg, woodruff, camomile flower, rosemary, hyssop, thyme, lemon mint, ginger, and basil. Usually three or more of these are used in combination. It is difficult to recommend a formula that is to everyone's liking. One authority on spiced meads suggests testing concoctions by steeping them in water or water to which one adds 12 to 14 percent of a flavorless

This cork has been properly seated in the mead bottle; the surface of the cork should be under the lip of the bottle.

alcohol. Since most of the herbs and spices mentioned are quite strong only a pinch is needed for each gallon of honey and water.

Aging

All sound alcoholic beverages improve with age. Aging is a complex process involving many chemical and physical changes in a wine or mead. Because of the time and investment required in aging many people have attempted to speed it up. However, no one has been able to come up with a substitute for the time-tested method of aging a wine or mead in a cask or bottle.

In the author's experience, light honey wines are drinkable after a year or two, although those aged five to six years are much improved over the younger meads. Meads made with darker honeys can profitably age for an additional year or two. Fruit meads should be aged for a minimum of two years and those made with strong, harsh fruit juices, such as elderberry and wild grape, will be best if they are consumed after ten to fifteen years.

Whereas it is true that most chemical changes are speeded up by higher temperatures, experience indicates that wines age best (insofar as flavor is concerned) when aged at about 60° to 70° F. When the new mead is clear—that is, when the yeast cells have settled—it should be placed in wine-type bottles that are corked or sealed with crown caps. Old meads, like many old wines, may develop a sediment in the bottle; however, the sediment can be left in the bottle and it in no way adversely affects the flavor of the final product. Aging is usually done in cellars because cellar temperatures are more uniform. Also, cellars are dark and sunlight can have a bad effect on the color of a wine or mead. Sunlight may darken mead.

*To enjoy the bouquet fully, bury your nose in a tulip-shaped glass
containing a small amount of the mead.*

Bottling and Labeling

A mead that is properly bottled and labeled will be much more appealing than one that is not. Even the person who makes mead only for his friends will find that they will accept the wine with greater enjoyment if it is presented to them in a labeled wine-type bottle sealed with a cork or cap. The label should indicate the type of mead and the date it was made; some people add additional information for the education and amusement of their guests.

Tasting Honey Wine

Most persons who produce honey and/or mead do so for the sole purpose of making a product that will please the senses, either their own or those of friends. A good mead has many enjoyable qualities which should be savored; those who drink it should not be in a hurry. A good mead should be brilliantly clear. Commercial wines are often made clear through filtration and the use of materials such as bentonite, egg albumin, or shocking the mead with a heat or cold treatment. However, most meads will become clear naturally upon standing. It is preferable to give meads sufficient time to develop clarity by themselves. Mead should be served in a colorless glass and when held to the light should have a clear, clean appearance.

A good mead or wine has a pleasant bouquet. One may best assess the bouquet of a mead by burying one's nose in a tulip-shaped glass in which there is only a small quantity of mead.

In tasting an alcoholic beverage, which is done after judging its aromatic qualities, there are several considerations. First, of course, is the immediate sensation upon sipping a small amount of the liquid. Oftentimes, and depending upon the beverage, one will experience quite different sensations when taking a small mouthful versus a larger quantity. Then there is the aftertaste.

Mead makers, like winemakers, seldom produce a perfect beverage. Sometimes the immediate taste or the aftertaste does not match the bouquet. Sometimes a cloudy mead has a superior flavor to that of a clear mead (though not usually). There is always the challenge of improving one's product, or when one produces a mead with a fine taste, the challenge of duplicating it. Seasons vary, honeys vary, and the conditions under which the products are made vary. A successful meadmaker, like a successful beekeeper, is one who blends his knowledge with the artistry of the subject.

Further Reading

Beekeeping has fascinated men for centuries. As a result there have been more books written on bees and beekeeping than on any other insect and most of the animals men husband. The books listed below are a few the author considers best.

If the author were to recommend any single one of these, it would be Dr. C. C. Miller's *Fifty Years Among the Bees;* it is written from the point of view of colony management. Miller's *Forty Years Among the Bees* is almost as good. The last edition of *Fifty Years* was printed in 1920 and the book is not available except through secondhand book dealers. Miller was a doctor who gave up practice to produce comb honey in Illinois. Few men were such careful observers of bee behavior. Miller wrote extensively for the bee journals and even in his own time was recognized as one of the great American beekeepers. Miller's book is poorly indexed and many of the important points concerning colony management can be found only by thorough reading.

There are also several books on bees that treat very specific subject material in detail. Some of these are valuable additions to any beekeeper's library. Lastly, every beekeeper should take advantage of the mimeographs, circulars, bulletins and booklets published by the state colleges and agricultural experiment stations. These are usually sold for a small fee or are sent free of charge. Often certain bulletins written decades ago still have application today, though obviously many of the older publications are difficult to obtain.

A library is probably the beekeeper's most important tool. There is not much in the way of management, equipment, methods or gadgetry that has not been written about by someone somewhere. Digging out correct ideas and discarding wrong ones is often difficult; profiting by someone else's mistakes is the easiest way to prepare a good management program.

Dade, H. A. *Anatomy and Dissection of the Honeybee.* Kent, England: Bee Research Association Ltd., 1961. This book describes the anatomy and physiology of the honeybee mainly for the amateur and suggests tools and practical methods of dissection.

Free, John B. *Insect Pollination of Crops.* London and New York: Academic Press, 1970. A detailed text on pollination problems around the world.

Frisch, Karl von. *The Dance Language and Orientation of Bees.* Cambridge, Mass.: The Belknap Press of Harvard University Press, 1967. This text summarizes the research von Frisch and his students have done on bee behavior.

————. *Bees, Their Vision, Chemical Senses, and Language.* Ithaca, N.Y.: Cornell University Press, 1950. Revised, 1971. A concise account of the basic experiments that led to our present understanding of honeybee behavior.

Laidlaw, H. H. and J. E. Eckert. *Queen Rearing.* Berkeley, Calif.: University of California Press, 1962. An excellent account of one of the more complicated aspects of apiculture.

Lindauer, Martin. *Communication Among Social Bees.* Cambridge, Mass.: Harvard University Press, 1961. This book contains many interesting and educational facts on bee communication.

Lovell, Harvey B. *Honey Plants Manual.* Medina, Ohio: A. I. Root Co., 1956. A small economical paperback book that describes most of the important honey- and pollen-producing plants in North America.

Ribbands, R. *Behavior and Social Life of Honeybees.* London, England: Bee Research Association Ltd., 1953. An excellent scientific textbook on the life and habits of the honeybee.

Smith, F. G. *Beekeeping in the Tropics.* Bristol, England: Western Printing Service Ltd., 1960. A modern book on beekeeping in tropical Africa.

Snodgrass, R. E. *Anatomy of the Honeybee.* Ithaca, N.Y.: Comstock Publishing Co., Cornell University, 1956. A detailed account of honeybee anatomy.

Index

Kelley, W. T., 142

labeling, 84
Laidlaw, H. H., 142, 218
Langstroth hive, 21, 23, 39
Langstroth, L. L., 20–25
life history, 184
life history of the colony,
    191–193
Lindauer, M., 218
locust, 165–166
Lovell, H., 219

mating, 131–134
mead, 203
metheglin, 203–204
methyl bromide, 119
mice, 102–103, 115–116
migratory beekeeping, 89–90,
    150–153
Miller, C. C., Dr., 66, 217
mites, 121–123
moving bees, 89–90

natural comb, 44–45
natural mating, 133
nosema disease, 111, 113–114

odor, 195
orange, 166–167
osmophilic yeasts, 105
osmotic pressure, 105
outbuildings, 34

package bees, 35–42
packing bees, 98–102
painting hives, 31
paradichlorobenzene, 119
pentachlorophenol, 25, 28
pesticides, 123–126
pheromones, 134–135, 196–199
Phillips, E. F., 160
pollen, 19, 156
pollen substitutes, 56–57
pollen supplements, 56–57
pollination, 11, 13, 172–181, 218
Porter bee escape, 75
privet honey, 171
propionic anhydride, 77
propolis, 29, 50, 57, 156
Pure Food and Drug
    Administration, 120
Pure Food and Drug Law, 143
purple loosestrife, 15

queen cells, 60, 140
queen cups, 60
queen excluders, 70–72, 88
queen mating nuclei, 139
queen rearing, 137–142, 218
queen substance, 128–134, 197
queens, life history of, 188–191

races of bees, 49–50
raspberry, 167–168
removing honey, 74–78
requeening, 92–93
reserve queens, 135–136